Patent Strategies for Business

Third Edition

Books by Stephen C. Glazier

Patent Strategies for Business
Third Edition

e-Patent Strategies
for
Software, e-Commerce, the Internet,
Telecom Services, and Financial Services
(with Case Studies and Forecasts)

Technology Deals
(to be published in 2003)

Patent Strategies for Business

Third Edition

Stephen C. Glazier
Washington, D.C.

LBI
Law & Business Institute
Washington, D.C.

Published by
LBI Law & Business Institute

Distributed to Booksellers and Libraries by
Ingram Book Co.

Parts of this book have been translated and published in Japanese, with permission, in the *Journal of the Japanese Institute of International Business Law*.

This entire book has been translated and published in Chinese, with permission, by the Peking University Press, Beijing, China, ISBN 7-301-01195-4/D.109.

Cataloging in Publication Data

Glazier, Stephen C.
Patent Strategies for Business/Stephen C. Glazier.--3rd. ed.
Includes index.
ISBN 0-9661437-9-5 (hardbound), 0-9661437-6-0 (softbound)
1. Patents. 2. Inventions. 3. Software-Protection. 4. Internet.
I. Title.

Library of Congress Catalog Card Number: 97-97126
ISBN: 0-9661437-9-5 (hardbound), 0-9661437-6-0 (softbound)

Printed in the United States of America. Hardbound and softbound.
Fourth Printing.

Summary Contents

Detailed Contents

About the Author

Stephen C. Glazier practices law in Washington, D.C., regarding patents, copyrights, trademarks, patent due diligence for M & A, patent demand letters and patent litigation, venture capital, licensing, and related business transactions for technology companies. He has a B.S. and an M.S. from the Massachusetts Institute of Technology (M.I.T.), and a J.D. from the University of Texas. He is a member of the bar in New York, California, Texas, and the District of Columbia, and he is a registered patent attorney with the U.S. Patent and Trademark Office.

Mr. Glazier holds six U.S. patents, in which he invents around prior art patents, for clients.

Mr. Glazier is the author of three books. He lectures frequently on legal and business topics, and is a contributor on legal subjects to the editorial page of *The Wall Street Journal* and other publications.

Mr. Glazier is a partner in the law firm of Kirkpatrick & Lockhart LLP, and is head of the Washington Patent Group of that firm. He may be contacted in Washington, D.C. at e-mail: glazier@alum.MIT.edu or sglazier@kl.com, telephone: 202-778-9000, fax: 202-778-9100.

Kirkpatrick & Lockhart LLP is a general practice law firm with about 700 attorneys in 10 offices, including about 45 intellectual property attorneys. For more information, see www.kl.com.

TO ORDER THIS BOOK

Patent Strategies
for
Business
Third Edition

It is easy to order a copy of this book (hardbound or softbound), or the Companion Volumes to this book: ***e-Patent Strategies for Software, e-Commerce, the Internet, Telecom Services, Financial Services, and Business Methods***, and ***Technology Deals*** (to be published in 2003). All you have to do is one of the following:

(1) Go to the Web, to
Amazon.com
Barnesandnoble.com
Booksamillion.com
or to one of the many other book selling Web sites.

(2) See your bookstore. They may have the book in stock, and if not, they can order it for you.

(3) Booksellers and libraries may contact their account representative at the Ingram Book Co., America's largest independent book distributor. Ingram distributes all of the publications of LBI Law and Business Institute.

Preface, Third Edition

Patents Are Business Tools

"Patents add the fuel of interest to the spark of genius."
-Abraham Lincoln

Popular demand for the second edition of this book encouraged us to reprint the second edition in an improved format with expanded distribution and a lower price. However, in working on the re-printing, we could not resist making additions and revisions. The changes quickly got out of hand, and as a result, the volume in your hands is actually now the third edition.

The first twenty-five chapters are largely unchanged, although there are many small improvements, updates, additions, and corrections. Particularly interesting changes were made in Chapter 3 ("Invent Around Your Competitor's Patent, and the Antidote"). Chapter 13 ("New Developments 1995") should now be read in light of Chapter 31 ("New Developments: 1996 and Early 1997"), particularly regarding the Patent Office software patent guidelines. And because of their importance, the PTO's new software patent guidelines are added in their entirety as Appendix 3.

All of Section VIII, beginning at Chapter 26, is totally new.

Again, let us emphasize that this book is not a "dumbed-down" or "beginner's introduction" to patent law. Instead, it is a sophisticated discussion of selected key approaches to make intellectual property serve a company's business plan and goals. The targeted reader of this book is a CEO or general counsel of a US or foreign corporation, whether large or small. Patent lawyers have also found this book useful to clarify a client oriented view of the patent profession.

A word of caution: Nothing is quite as simple as it first sounds. Of necessity, many complicated legal technicalities are glossed over or not discussed in this book, in an effort to communicate fundamental strategies. These strategies are valuable in many cases, but may be unavailable in other cases. Any such unavailability may arise from special facts, surprises, financial constraints, changes in the law, or otherwise. (Also, any part of this book may be obsolete or otherwise in error at the time you read it.) And, of course, it seems that any useful legal statement, and certainly any evaluation or judgment of merit, is arguable. So please consult legal counsel, your special facts, and current law, before assuming that anything herein may work or be advisable in any particular case. (But let's not lose the forest because a few trees change colors; it is our expectation that the big ideas in this book will survive many changes in the details of law and business.)

The reader is encouraged to communicate to the author any comments regarding preferred additional topics for the fourth edition of this book. Thank you for your interest.

Stephen C. Glazier
Washington, D.C.
2 September 1997

Patents Are Business Tools

"Congress shall have power... To promote the Progress of Science and useful Arts, by securing for limited Times to Authors and Inventors the exclusive Right to their respective Writings and Discoveries."

U.S. Constitution, Article I, Section 8

This book is written as a practical guide to the use of patents as effective business tools. That is, this book is written for businessmen and attorneys who are not intellectual property specialists, but who do have opportunities that can be pursued by practical patent strategies. Other areas of intellectual property law are also touched upon, particularly regarding copyright and trade secrets, where they apply to software.

This book is designed to be a tool for the reader, to present ideas and strategies that have been successful for others, and that will be successful again. This book is not an academic treatise or survey of the entirety of patent law for the patent attorney specialist. It should, however, be useful for the patent attorney, since the book explicitly discusses important strategies and concepts that are seldom analyzed in print. Current legal developments and their practical applications are also discussed.

The approaches in this book have grown out of specific transactions with specific clients, so we are confident that the ideas in this book have passed at least their trial runs in the real world. However, we have been careful to keep all the discussions here on a generic level, so that no confidences regarding any particular party are revealed. Any references to specific companies or individuals in this book communicate only information that is disclosed in the public record, and which involves parties that this author has never personally represented.

There are several major trends in patents in the last 10 years that have revolutionized the patent business, and which we expect to continue for the foreseeable future.

Patents were made more enforceable, and appellate law uniform and more correct by the creation of the Federal Circuit Court of Appeals in the mid-'80s. This arose out of a general recognition that the U.S. was in a competitive world marketplace that it could no longer dominate, and hence the U.S. had to protect its intellectual property in order to remain competitive. This economic situation will not change, so we can count on this legislative change in patent law to be permanent.

Also, software and computer applications have been determined to be patentable technology, if they are new and not obvious. This legal development takes place as software becomes more prevalent in daily life, and as custom circuits are being replaced as a design approach in favor of standard chips with custom programming. At the same time, hardware becomes cheaper, for processing and for memory. The result is that in more and more industries, more of the innovation is in the form of patentable software applications. This has lead to the shocking application of patents to new areas such as telecommunication services and financial investments. Also, we may be about to see the rise of business method patents, to a par with the current patentable status of all other man-made processes and methods. At the least, we can expect for software patents to become a factor in most industries in this decade.

Patents are not just for gizmos anymore.

We are also seeing the globalization and harmonization of patent law across our planet. This has lead to the currently pending amendments to the U.S. intellectual property statutes that arose from the GATT treaty, and these amendments look minor compared to what we may see in the next five years. This trend has also lead us to spend more time in Munich (at the European Patent Office) and Tokyo (at the Japanese Patent Office) than we ever would have guessed, and to the fact that about half of our clients are foreign to the U.S.

These developments have greatly increased the power and value of patents. Consequently, patent questions are now critical in many commercial and financial transactions. Judgments for patent infringement are bigger (Kodak recently paid Polaroid almost $1 billion for patent infringement, and Litton more recently received a judgment against Honeywell for $1.2 billion), and patents are more often enforced when litigated. It has also been reported that Texas Instruments has received more revenues in some recent years from its patent royalties than from the sale of its products. This has put more pressure on management to develop their patent portfolios, and to avoid infringement. It has also put pressure on Congress to fix the ownership section of U.S. patent statutes (see the appendices herein for one proposed bill).

As a result, financiers are now beginning to do real patent due diligence, and are looking for patent attorneys who can do a deal.

Also, with the possibility of triple damages for patent infringement, we have seen the beginnings of a plaintiff's contingency bar for patent infringement, and an increasing ability of small patent owners to negotiate licenses with major corporations. Also, we have seen the beginning of an evolution away from the patent specialty law firm, to the patent group at a general practice law firm. We have also seen the first younger transactional patent attorneys.

With the first appearance of biotechnology companies, the courts have agreed that man-made life forms are patentable. (This development is not dealt with in this edition of this book, and is a matter for the third edition.)

We have also seen that copyright can be used to obtain some protection for software. However, copyright protection is weak compared to patent protection for software (when you can get a patent). The unique overlapping nature for copyrights and patent rights for software require that both sides of this ownership coin be attended to for software.

We have also seen case law develop and clarify the legal side of inventing around a competitor's patent (a process of legal development which is still continuing). This makes the invent around process more reliable and practical. Also, this in part enables what we personally find to be the most exciting innovation of this book, which we call the Rules of Virtual Genius.

Inventing and patenting are two different things that are part of a larger commercial effort, to make money from new products and services. Surprisingly, current developments in technology and law are actually working in the same direction, so that good inventing can facilitate good patenting, and, most unexpectedly, good patenting can facilitate good inventing. We will probably never be able to advise, in general, to go to a patent lawyer for an invention (instead of with an invention); however, inventing and patenting is a lot easier than most technical people realize, especially if they stay focused on the Rules of Virtual Genius.

We also have found that the Rules of Virtual Genius are teachable to most business and technical people, and lessons on this subject can inspire profitable innovation.

We also have seen since the late '80s that applications at the U.S. Patent Office have been growing at about 15% per year, while software applications have been increasing at about 30% per year.

In effect, this is the book (or at least part of it) that I wish someone else had written for me when I was an engineering student in the '60s at M.I.T., or a law student in the '70s. Hopefully, it can now give others a faster start and a better direction in the '90s.

The point of this book is to help businesses make money. It does this by pointing out patent strategies and new legal developments that may offer profitable opportunities, in specific circumstances, within a company's budget and schedule.

Stephen C. Glazier
Washington, D.C.
23 October 1995

WHAT THE REVIEWERS ARE SAYING ABOUT THIS BOOK

Patent Strategies
for
Business
Third Edition

"Stephen Glazier's book entitled *Patent Strategies for Business, Third Edition*, is a field manual for the intellectual property strategist to start thinking and acting... Glazier's book is one of the few sources which makes the effort to approach the patent field as a matter of strategy rather than as a matter of... how the authorities line up on each particular legal issue... Glazier's book lets the reader understand in a brief and manageable way how things work in the patent field... The writer wishes Glazier would convert his book into a multi-volume loose leaf series for which there is surely strong need and probably no better potential author or editor."

-*Intellectual Property Rights News*
Volume 2, Number 3
(Winter 1998)

WHAT THE REVIEWERS ARE SAYING ABOUT THE COMPANION VOLUME

e-Patent Strategies

for

Software, e-Commerce, the Internet, Telecom Services, Financial Services, and Business Methods (with Case Studies and Forecasts)

"Mr. Glazier focuses on practical strategies for business-people who use patents to advance their commercial goals. The book is refreshingly different from works about patent law that are more useful to lawyers than businesses. Complete with specific ideas on what to do to improve profitability, this book will be a welcome addition to many collections... This is an excellent book. Good for both managers and patent lawyers."

-the *Amazon.com* review
May 1, 2000

Acknowledgments

My clients over the years have pressed me to invent and organize the ideas found in this book. Ultimately, those clients gave these ideas their greatest seal of approval: they used them. I would like to give my clients, past and present, special thanks.

(However, I would like to tell Liz, Karen, Bill, Joe, John, Tom, Barry, Paul, Gerry, Ted, Jim, Mike, Gene, Tim, Heinrich, Anders, Dexter, Jud, and Chuji that they can relax. My best ideas remain in my unpublished confidential correspondence to them.)

1

Strategic Management of Intellectual Property Assets

"Patents ... have the attributes of ... property."
35 U.S.C. 261

"File early. File often."

For fiscal year 1990, Texas Instruments Corp. reported that its revenues from patent royalties exceeded its revenues from manufacturing. This is significant in light of the fact that Texas Instruments considers itself a manufacturing company, not a contract research laboratory. Apparently, Texas Instruments has found that a program for managing intellectual property (which includes patents, copyrights, trade secrets, confidential information, and trademarks) can prove to be a cashflowing profit center. A similar opportunity may be available to many other companies, in a surprisingly wide variety of industries.

Five Goals

A company's intellectual property management program should include one or more of the five following goals:

1. Protection of a Company's Products, Services, and Income. An obvious example is that a patent can be used to suppress competition, because patents grant the exclusive use of an invention to the patent holder for a specified time, and patents can be actively enforced against infringement through the courts.

This goal is driven by a company's desire to protect its share of the market, and it focuses on a company's current and near-term products, services, and customers. Companies interested in using intellectual properties in this way should review their business lines to find opportunities for patent monopolies.

2. Generating Cash by Assigning (Selling) or Licensing (Renting) Patent Rights to Others. A company may decide, as in the example of Texas Instruments, that it does not want to develop a certain technology, but realizes it can make money by licensing patent rights on that technology to another company.

3. Obtaining a Legitimate Monopoly for Future Exploitation. The point of this goal is to put new technology "on hold" until a decision is made about how it can best be used.

4. Protecting Research and Development Investments. This goal is a form of insurance. If $1 million has been spent on an R&D project, for example, and obtaining a patent on the technology costs between $10,000 and $20,000, spending the money is a good investment because a patent monopolizes title to the technology until its value can be accurately determined.

5. Creating Bargaining Chips. A company sometimes develops a large patent portfolio to be used as "chips" to negotiate cross-licenses with potential competitors that may, in the future, claim that the company is infringing on the competitor's patents. Again, the motive is insurance, that is, if a company is concerned that another company may sue it for patent infringement, it may wish to trade its rights in its own broad patents for rights to a competitor's broad patents.

Accumulating as many patents as possible should *not* be one of the goals of an intellectual property management program. By itself, a patent may have no value. Unless it can be used as a tool to pursue the larger business interests of a company, a patent only accrues costs.

A Nine Step Program

Developing a strategic intellectual property management program can be accomplished in nine basic steps. The following discussion focuses on patents, but analogous steps apply to copyrights, trade secrets, confidential information, and trademarks.

1. Obtain Disclosure of Inventions. One effective way for some companies to encourage employees or consultants to disclose their ideas for inventions is to offer a program of cash incentives. This is typically a one-time payment or a regularly paid percentage of the income resulting from an invention. In some companies, patent disclosure forms are distributed periodically as a way of soliciting useful ideas regarding inventions.

Another effective method has been for patent counsel to meet with a company's technical people to ferret out together innovations that may yield patents of value in the marketplace. It can be particularly useful to do this with a focus on a new product or service just before its market introduction. With companies with a particularly intense product development schedule, scheduled regular monthly meetings of this sort can yield good results in identifying important patent opportunities.

2. Review the Disclosed Ideas. This is done by a review board that typically includes manufacturing, R&D, marketing, and finance personnel. The board may decide to patent an invention immediately, keep it a trade secret and develop it further, or disclose it to the public domain without patenting it (an act that preempts patenting by others).

3. Establish a Confidentiality Program. Staff must be trained in confidentiality procedures. Otherwise, normal business practices, such as advertising and press releases, will very likely allow ownership of new inventions to lapse into the public domain by default, before their exclusive ownership can be acquired by registering patents and copyrights. Even worse, title may come to rest in the hands of employees and consultants, who may then exercise exclusive rights in competition with the company.

A confidentiality program should contain the following types of items, adapted to a company's individual circumstances:

- Signed confidentiality and noncompetition agreements with all employees, requiring that confidentiality be maintained during and after employment, and stating that employees are restricted from competition with the company.

- A written confidentiality policy document signed by all employees, indicating their knowledge of and agreement with the policy.

- A review of all publications and speeches by company personnel before such presentations are made public.

- Use of dated laboratory or research notes by all company technical personnel.

- Labeling of all trade secret documents as "confidential," the use of paper shredders for trade-secret trash, locks and other security measures to protect secret information, computer and fax passwords, copy-protects, and phone-line scramblers for modem and fax lines.

- Exit interviews with all departing employees reminding them of their post-employment confidentiality obligations.

4. Establish a Licensing Program. A licensing program works best when it is regarded as an independent profit center whose staff is compensated with a percentage of the income they obtain through the licensing of patents, rather than when the program is administered by a staff on salaries. The key here is individual incentive generated by performance compensation. It is also productive to combine this with the authority to close a transaction, or at least an environment in which a practical decision can be approved.

5. Establish an Enforcement Function. This step attempts to ensure that no one infringes the company's patents, and it usually requires constant policing and monitoring of the market in order to challenge infringing products and services, and invalidly issued patents owned by others. Foreign patent applications should also be monitored, and, where necessary, companies can initiate opposition proceedings in foreign patent offices.

An enforcement function should also include a method of negotiation so that the company is able to suppress infringement without having to engage in litigation. In tough cases, however, companies must be prepared to litigate against infringers both in the United States and in foreign countries.

6. Avoid Infringement of Patents of Others. Before a company spends substantial money to develop and market new products or services, it should make sure that its ideas do not infringe the patents of other companies. Current law encourages companies to obtain a written opinion regarding noninfringement by competent patent counsel not on the staff of the company. This document could save a company from paying treble damages for infringement.

Note that it is particularly difficult to do reliable prior art searches for software developments, although this situation is slowly improving. (This is largely because documentation of software is traditionally sketchy, there is only a small history of software patents, and there is no standardized terminology for many software developments. However, the written public record of prior art is expanding.)

7. Monitor the Patent Activities of Others. This includes companies in the United States and abroad. Such monitoring is an excellent way to obtain information about the operations of competitors and to keep up-to-date on technologies of interest to one's own company.

8. Ensure that Title to New Developments Comes to Rest with the Company. Contracts should be signed by all employees and company consultants making it clear that all rights to development are the property of, and shall be assigned to, the company. Similar agreements must be obtained from all third-party business associates, including joint-venture partners and suppliers.

9. Determine in Which Foreign Jurisdictions Counterpart Patent Applications Will Be Filed. A U.S. patent application yields only a U.S. patent, which gives rights only within the jurisdiction of the United States. If a company is interested in marketing a device internationally, it should decide in which foreign countries counterpart applications should be filed.

Lobbying Required

In addition to the steps outlined above, companies may find it worth their while to participate in lobbying efforts to remedy some of the deficiencies in current U.S. patent law. For example, U.S. patents, trademarks, and copyrights need a complete federal system of publicly recorded ownership. Presently, it is impossible to perform a good title search for intellectual property in the U.S., and there is no clear statutory provision for the procedures for using intellectual property as collateral for loans and obligations. Also, U.S. patent law needs to be harmonized with international patent law. Important steps would include adopting a first-to-file rule for awarding patents, publishing information about patents while they are pending, and permitting third parties to challenge pending patent applications. Enacting these measures would go a long way toward bringing U.S. patents in line with world standards, and making them more reliable and valuable.

The Appendices herein include an excellent bill that has been proposed to Congress to correct some of these problems. The active support of this bill in Congress by members of the technology community is encouraged.

2

A Corporation Can Invent-Just-In-Time: The Rules of Virtual Genius

Virtual: *Existing or resulting in essence or effect though not in actual fact, form, or name. (From Latin virtus, meaning capacity.)*

Genius: *Exceptional or transcendent intellectual and creative power. "True genius rearranges old material in a way never seen . . . before." (John Hersey) (From Latin genius, meaning the deity of generation and birth.)*

from the American Heritage Dictionary of the English Language.

Virtual Genius: *Someone who is not an actual genius, but who can invent and patent like one by following the Rules of Virtual Genius.*

Technical innovation, inventing, and the patent process are essential to the competitive health of many corporations. Furthermore, with the expansion of the subject matter of patents to include software, financial products, communication services, smart equipment, and business methods, the issue of inventing/patenting is important to many industries that traditionally are not patent conscious. An urgency is added to this because of the possibility of triple damages for patent infringement, the appearance of a contingency fee plaintiff's bar for patent infringement, and the development of personal liability of officers and directors for corporate infringement (liability that can be civil and criminal).

The need to manage corporate invention and the development of a corporate patent portfolio is hampered by a false view that invention is the result of a flash of creative insight striking a genius in R&D like a lighting bolt. This metaphor sets up a false paradigm of corporate patenting that would undermine the rational management of the development of a corporate patent portfolio.

Fortunately, however, the reality can be different from this false paradigm. The corporate patent portfolio can be rationally developed and exploited. Spontaneous creative insight can be valuable, but valuable patents are more likely if invented-on-demand, as it were, by normal corporate employees following the basic rules of profitable invention.

Inventing and patenting is often easier than generally thought to be the case, especially if certain rules are known and followed. Most people seem to think that patents are for someone like Einstein or Edison, and involve great scientific discoveries or great technical leaps forward. Although this is occasionally the case, most valuable patents involve no great scientific discovery, and represent only a modest advance in fundamental technology, if any, and are made by ordinary educated people working in a corporate environment.

Furthermore, patents are not the point of most valuable patents. In fact, invention is not the point of most valuable patents. Instead, most patents are obtained for the proper business purpose of keeping competitors away from the market for a new product or service. That is, most patents are not technology driven, and they are not the result of great new technology. Instead, most patents are product driven. That is, when a new product or service is planned, the providers want to inhibit competition. Therefore, the providers adopt a strategy to fight the competition that they know will develop. If possible, the strategy includes intellectual property strategies, which in turn include patent strategies, if the providers are so fortunate. The patents are then developed to protect the product or service in the market.

Seen in this way, the invention team should include the entire product development team, including sales, manufacturing, design, R&D, finance, and legal participation.

With this orientation in mind, the invention and patenting process can be facilitated by following the technical and legal rules below. And it can be a lot easier than most people think, if the right focus is maintained.

The Hurdle to Patenting

The legal standards for patenting are lower than commonly realized. Although the legal standards for patenting are arcane, and unfortunately it probably does take a patent lawyer to apply the standards, it is a fair summary to say that the hurdle to a patent often centers around the concept of obviousness. The innovation in a patent application must be, legally speaking, not obvious in light of prior practice at the time the application is filed, in order to yield a patent.

To most technical people, this obviousness hurdle seems to mean that, in retrospect, an invention must involve a new scientific discovery, or a new engineering principle. This, however, is not the legal view of obviousness. Legally speaking, hindsight cannot be used to find obviousness for a new idea, and hence hindsight can not be used to deny a patent. Also, a mere extension of well known engineering principles to give a new but unsurprising solution to a problem can be, legally speaking, not obvious and quite patentable. Indeed, in some cases, the fact that an invention could have been made much earlier than it was, is by itself a proof of non-obviousness, and hence patentability.

The obviousness question may also be misconstrued by the non-patent lawyer public because a very easy solution to a new problem is not necessarily obvious in the view of patent law. (This concept often comes into play since many patentable solutions come quite easily once the problem that they solve is clearly understood, especially when the problem is new. In retrospect, these inventions are more the result of practical analysis of new problems, than the result of genius and profound insight.) The amount of time and money spent, or the lack of it, in inventing is not interesting to the legal question of patentability.

For software driven invention, the key to successful innovation is often in determining what the software should do, that is, what functions should the software offer to the user. The key to innovation in software driven products and services is often not in how the functions are delivered. In other words, once it is decided what the user in the marketplace most wants, getting the source code written and de-bugged, and assembling the hardware platform to execute it all, can often be routine. Put another way, the novelty in software patents often comes more from the novelty of the perceived problem than from any surprising solution to the novel problem.

Inventing Around as a Response Strategy

One response to a problem patent owned by a competitor that threatens to keep your company out of a lucrative market, is to invent around the competitor's patent. That is, your company can develop a product, and possibly patent it, that does not infringe on the competitor's patent, yet still penetrates the same basic market and customer base.

The "invent around" strategy is a bona fide response to threats of patent infringement claims by competitors, and is one of several strategies that might be simultaneously pursued in response to a letter or other notification from patent holders to "cease and desist" from infringement. As it turns out, the rules to invent around a competitor's patent are largely the same as the rules to invent-just-in-time, or to invent-on-demand. Conceptually, invention-on-demand, or invention-just-in-time, can be thought of as inventing around the whole universe of prior art, rather than just one competitor's patent portfolio. The target of invention-on-demand may be larger than inventing around, but the processes are about the same.

An Engineering/Legal Nexus

The effort to invent around a competitor's patent is partly a legal effort, and partly an engineering effort. The "invent around" effort requires a measure of creativity. It takes place at the nexus of law and engineering, and can best be pursued by individuals or a small team with a creative track record and professional experience in both law and technology. In a specific case, this effort may be largely a question of creative legal draftsmanship and argumentation, with a seemingly modest amount of creative engineering.

Invention is not usually a mysterious process involving a genius waiting for lightning to strike. Invention can also be tractable and manageable, within limits. The successful "invent around" project is a matter of professional knowledge, focused on the proper values, with a minimum of creativity required. Normal individuals with the right knowledge and focus can "invent on demand," without a lot of genius. Normal people in a corporation, if they follow the basic rules described below, can get the same results of prolific and profitable patenting that are usually associated in popular imagery with geniuses. Think of these rules as the Rules of Virtual Genius. A corporation of normal people can perform like geniuses, even when they are not. They can be Virtual Geniuses.

In fact, most successful patents are probably more related to product development efforts, using known scientific principles, than to basic research and development. The immediate motivation of the "design around" or "invent around" strategy is often to distinguish a new product for immediate sale that avoids infringement of a specific patent. However, the same approach, when applied against the entire body of prior art in a field, can lead to larger and more profound steps forward, that can create new industries.

Twelve Rules of Virtual Genius

There are basic rules to "invent-just-in-time," or to "invent-on-demand." These rules are aimed at inventing around a competitor's patent, or developing a new patentable product that leapfrogs the marketplace. These rules are the Rules for Virtual Genius.

The Rules of Virtual Genius are teachable, and are useful to focus and inspire invention and the patent process.

Rule 1. Eliminate a Part. The primary secret to inventing on demand is to carefully inspect the competitor's patent and the prior art in the field, and then to invent by eliminating the non-essential elements that were previously thought to be essential. This often requires realizing a new design concept, and changing some of the remaining parts. This approach provides both a legal and an engineering advantage.

On the legal side, elimination of an element that is non-essential for engineering purposes, yet which is claimed in the independent claims of the competitor's patent, can be a way to avoid the legal test for infringement. The basic legal test for infringement is that one of your products uses every element claimed in an independent claim of your competitor's patent. If you manage to develop a way in which the product may be made without one of the claimed essential elements of your competitor, then you may avoid the legal test for infringement.

On the engineering side, the elimination of a part usually results in a product that is cheaper to make, and is more reliable.

The elimination of parts is a deceptively simple goal, but it is at the core of the invent-around project. Seldom do legal and engineering values work together to encourage the same behavior; but with the elimination-of-parts rule, patent law and engineering synergistically cooperate to direct invention with the common guideline to seek elimination of parts that were previously thought to be essential. This tends to yield both a better product and a more patentable invention. (Of course, in specific cases, there can be other ways to improve a product and to get a patent. But, any time your company can eliminate a part, your company has achieved something notable.)

It is striking how often a close analysis of the independent claims of a competitor's patent indicates elements which are not strictly essential from the customer's point of view, or which can be rendered non-essential by a relatively low level of inventive insight. Looking at the same issue from the other side of the coin, it is often striking how a minimum of creative engineering insight can allow a product to be re-engineered with fewer parts, once the reduction of parts is conceived of as an explicit goal. When it comes to inventions and patents, less is better and more is worse.

Where a part cannot be found to eliminate in a competitor's patent, try to find a part that can be structurally changed to the point that it is a different part, even if it may function the same.

If a part cannot be eliminated or structurally changed, then try to find a part that can be changed to function differently, even though it may be the same structurally.

In the case of trying to invent around a competitor's patent that has claims written in the so-called "means plus function" format (that is, the elements recite "means for [doing a specific function]") another alternative is presented. The patent statute limits the meaning of these claims to the structures described in the accompanying patent specification, and their equivalents (see 35 U.S.C. 112, paragraph 6). To invent around claims in this format, first try to eliminate an entire element, as discussed above. If this cannot be done, identify a structure that can execute a function (that is, that can be one of the "means for [a specific function]") but which is a structure that is not itemized in the specification of the patent. Then claim your own invention using that specific structure.

In the case of software driven inventions, this rule can be straightforward. Software elements in claims are usually written in means-plus-function format, where each element amounts to a box in a macro-level flow chart of the software. To invent around the software elements of a claim, change the flow chart to have less boxes or different boxes, but still result in a product or service that penetrates the same market niche. (The trick in drafting and inventing software patents is to make the claims broad enough and simple enough so that they cannot be simplified or changed, and still result in a product or service that penetrates the original market niche of interest. That is, try to invent and patent in a way that is hard to patent around.)

Rule 2. Do Not Add Parts. A related rule for inventing on demand and designing around a competitor's patent, deals with analyzing problems in early prototypes. The natural human tendency seems to be to address problems in a product by adding parts and functions. However, it is often better to address problems in prototypes by eliminating or changing parts, not by adding parts. This often takes some minimum level of creative insight; however, it is surprising how often a normal engineer can do this once he or she consciously recognizes this goal. It is relatively easy to solve a problem by adding new parts, but it requires a new concept to solve a problem by eliminating or merely changing a part. This rule can be summarized with a slogan of "more concept, less gizmo" or with the phrase "design with an eraser, not a pencil."

Rule 3. Use a Lean Broad Design Team. This goal of leaner design is best addressed in most corporate cultures with a lean design team. It seems to be an axiom of human organizations that the larger a committee is, the less able it is to develop a focused innovative design. We have all heard that a camel is a race horse designed by a committee. It was a very large committee. (The problem here is that in a group, agreement is purchased by allowing each individual to make a contribution to the design. Hence, in a large group, you have too many contributions that are politically required, with the result being a design which is too complicated.)

A design team needs input from manufacturing, design, sales, finance, law and management. In a large corporation it is better to have on a project team of, for example, six people, one from each of these areas, than to have six technical people from R&D. Breadth and synergy is more creative than depth and a negotiated settlement on each technical point.

Rule 4. Focus the Product. The corollary to Rule 3 is to design a product with as narrow a target function as possible. It is possible to design a device that does one thing very well; however, it is often impossible to design a device that does many things very well. You might call this the Swiss Army knife rule. Unless you are in the toy business, don't design products like the Swiss Army knife, which tries to do too much to do anything well.

In discussing invention in general here, we often talk about the more specific project of inventing around a competitor's patent. These are related activities, of course, since inventing in its broadest sense can be viewed as inventing around everything that went before. In this light, inventing around your competition's patent is just a more focused and immediate application of the broader inventing project.

In inventing, and in inventing around a competitor, we really have two goals. We want a result that is patentable, so that we can keep our competition from copying it for the term of the patent, and we want to avoid all the patents of others, so that we can practice our patented invention without being stopped by other earlier patent holders.

Examples

These rules of invention can be seen at work in many developing lines of U.S. patents. For example, see the development of the breath activated medication spray. This is a long sought device to activate a measured dose inhaler ("MDI") of aerosol medicine to be inhaled. The idea is not to manually activate the MDI, but to automatically activate it by inhaling. For medical reasons, breath activation works better. Many U.S. patents describe such a device, including No. 3,636,949, No. 3,789,843, and No. 4,648,393. These show a variety of devices which are expensive and complicated, and never marketed. Compare these with the later design in No. 4,955,371, by Zamba et al. This design has eliminated parts to the extent that it has only one moving part. Therefore, it is cheap, easy to use, and disposable. '371 has been authorized for sale by the FDA, and is expected to be marketed soon. Comparing No. 4,955,371, by Gene Zamba, with the previous patents in the line, clearly shows the engineering and legal advantages achieved by applying a new design concept to eliminate parts previously thought to be essential.

After reviewing the first four rules of invention above, one might ask where to look to get some of the best ideas for inventions. The following additional rules point to some of the most fertile areas today.

Rule 5. Exploit Components with New Low Prices. Look for components that recently have become dramatically cheaper. Whenever this happens to an item, the item newly becomes a potential practical replacement for other items, or an addition to other items. This activity can constitute a patentable invention, and a good new product.

One example of this is the micro-processor chip. As these computers on a chip become cheaper, they become feasible to add to a wider array of previously existing equipment. (See Rule 6 below.) Another example is the GPS (Global Positioning System) receiver. As these brilliant devices lose one or two zeros from their unit price, they open up whole areas of economic applications that were never anticipated by the original developers of what was at first just a navigational aid.

Rule 6. Make Old Equipment Smart. Put a computer chip and a keypad on just about anything and you have a platform for the new smart version of the thing. Then find out what the market would like the smart thing to do, and program it accordingly. This can give you a new patentable product.

The Pentagon did this with smart bombs. But it has also been done with smart electric motors, that schedule their work to concentrate power usage in off-peak hours when electricity is cheaper. It has been done with smart carburetors that can choke themselves when the engine is cold. It has been done for smart cellular phones that search out the microwave cells with the best reception to communicate in. The list seems endless and growing. And even if the new hardware configuration resulting from applying this rule is not patentable, the programmed smart hardware configuration often is.

Rule 7. Exploit New Communication Devices and Services. Examine any new communication device or service to see how it enables a further new communication service. Services may be patentable if they represent new processes for delivering the service, especially if some software is involved. In communication this is especially vital because the field is benefitting from an avalanche of new cheap devices that can allow new combinations of devices and services that can serve as the platforms for new services. The new pieces in this expanding tinker-toy set include: fax machines, fax modems, cable modems (with capacities of 10,000,000 bits per second), 900 numbers, data networks such as Internet, cheap multi-media PC's, cheap large hard drives, wireless phones of several types, satellite communication, intelligent agents, video-on-demand, cable access to data networks, cheap RAM, robot telephone operators, voice mail, E-mail, cable, fiber optics, LAN's and WAN's, pagers, beepers, smart pagers, GPS receivers, cheap effective encryption algorithms, and digital communications, to name just some of today's new pieces of the puzzle.

Different combinations of these pieces can enable new patentable communication services. For example, Video Jukebox Network, Inc. has broad patents on the basic concept of the video jukebox channel on cable television. This is a fundamentally new service that was enabled by a combination of cable networks, rock videos, PC's, and 900 numbers. There are many more examples, realized and pending.

Rule 8. Computerize a Previously Manual Process. Computer software and computer algorithms can be patentable. If your company is the first to computerize an old function or process, then your company may be able to patent the concept of the computerization. It may not seem like an act of Albert Einstein, but it can be quite a valuable patent, and such patents are being regularly issued to corporations. For example, Merrill Lynch has a patent on the computerized bookkeeping for its CMA account for checking/brokerage/credit cards.

Rule 9. Use New Materials. New applications of new materials in old devices can lead to superior performance, and patents. For example, the development of lightweight, strong, heat resistant composite fiber materials has led to surprising advances in airplanes, rockets, golf clubs, clothing (see, for example, bullet-proof clothing), and industrial abrasives (see, for example, diamond interlaced Kevlar sanding belts).

Using an old material in a new way can have similar results, but the opportunities may be fewer than for new materials.

Rule 10. Focus on the Software. Software is now clearly candidate subject matter for patents, and lots of software patents are being issued. This amazes many people (and annoys some), but now every new computer program should be considered for a patent, and analyzed for possible infringement of the patents of others.

Even parts of computer programs are patentable. That is, computer algorithms are patentable. Even a mathematical equation embodied in a computer algorithm for a certain purpose may be patentable. This is a striking new development in patent law that has probably opened up more new territory for possible patenting than any single development since Ben Franklin brought up patents and copyrights at the Constitutional Convention in Philadelphia over two centuries ago.

Computer software is patentable because it is a type of process (processes are patentable), because a computer programmed with software is considered a unique device for each program that it may run (devices are also patentable), and because a floppy disk with unique software on it is considered a unique article of manufacture (articles of manufacture are patentable).

Rule 11. For Software Only: Find New Functions. Software is different from other technologies. Often the uniqueness of the inventive effort in software, smart equipment, or software-hardware hybrids, is more in conceiving new functions for the software (the "what does it do"), than in conceiving how to make it do what it does.

For software, once you conceive of what you want it to do, then getting it done may be a straightforward programming project. All the invention may be in the RFP ("request for proposals"), written by your V.P. of Sales, after reviewing the responses to your last customer survey.

This is a radical departure from, say, the mechanical arts. Take the airplane, for example. For centuries, we tried to make a heavier than air flying machine. We knew the "what," but the "how" was very difficult.

But take, for example, one of the first famous software patents, the Merrill Lynch CMA patent (discussed elsewhere herein). Once Merrill Lynch decided what features they wanted the CMA account to have, they just arranged for programmers to instruct the computer to keep the books in that manner and to issue one combined statement at the end of the month reflecting checking, credit card and security brokerage activity. This was a routine implementation for an inventive service function.

(The "how" of software techniques for implementing software functions can, of course, also be patentable, and is indeed the subject of patents. For example, there are patents on how to process data for graphic images, in real time. However, these programming technique patents seem to be a relatively minor percentage of the body of software patents. I also suspect that they tend to be less valuable, in general, because they do not address a mass market product, or any particular product at all. They can also be hard to police.)

This leads us to the observation that the "less is more" attitude of Rules 1 through 4, above, of the Rules of Virtual Genius, may not be so critical for inventing software. Lean, focused source code can still be important in writing software, but in conceiving inventive new software products, new concepts of functions that directly connect to markets are more important. And as long as the combination of functions is new and market responsive, the combination may be quite complicated and "Swiss-Army-knife-like." The kitchen sink approach to new software functions can work, if it is market responsive, just as well as a tight focused approach to new software function.

The decreasing cost of microprocessing power and memory is a major contributor to this trend. At least in the PC world, for software, the trend is not "less is more." The trend is "too much ain't enough."

Also, note, that a new combination of old software functions may be just as patentable as a new combination of old mechanical parts.

Also note that a new software method of providing old software functions can be patentable, although the functions themselves can no longer be patentable.

Also note that providing old mechanical functions, or manual functions, by a new software application can be patentable. That is computerizing well-known activities can be a new patentable "how" for an old "what does it do."

Rule 12. Mind the Esthetics. One goal for a new product or service is a good look that adds nothing to a product's production cost, but dramatically increases its sales. A good look can be pure profit. And elements of non-functional ornamental appearance, if they are distinctive, can be protected by something called design patents (what we discuss elsewhere herein are utility patents, that is, functional patents), and trademarks. (Admittedly, this is a bit far afield from our focus on patents and technology, but these are difficult points to avoid in discussing new products. We should remember, too, that all intellectual property strategies are best used together, not in isolated subspecialties.)

A very effective way to generate good design in an environment of technical innovation is to let form follow function. That is, abandon preconceived ideas of what a new product should look like. Instead, let a new product look like what it is most functional and cheapest for it to look like, and then modify it a bit, if necessary and without making it cost more to produce, to make it look a bit nicer and recognizable. Let the appearance reflect the innovation, and not cover it up. With a new inventive product, this will yield a new distinctive appearance. This distinctive appearance will promote brand distinction, and will be associated with the superior performance of the new invention. Hence, the new appearance may look a bit weird to begin with, but will end up being the new standard of appearance.

The new appearance will also be more economic, because it will be just what the bare unadorned product looks like, without the expense of making it look the old way, that is, without the expense of making the new product fit into the old product's shell.

This principle is seen over and over again. For example, when Ford wanted a more fuel efficient car, they reduced wind drag by using airplane techniques for a smooth aerodynamic exterior. The resulting Taurus created the originally shocking "aero" look. The Taurus became the best selling car in America, and every car on the planet seems to have adopted the aero look, instead of the boxy angular look that dominated cars when the Taurus first appeared. Form followed function, and the funds followed Ford. (Unfortunately, the aero look, in general, was probably never patentable, although probably not exactly copyable for other legal reasons.)

A similar thing happened later at Chrysler. They wanted more room inside of a car, without making the car any bigger. Chrysler did this by changing the basic architecture of the car. They moved the wheels closer to the corners so that the cab inside of the car would be expanded, even though the outside dimensions of the car were not increased. The result was the very popular Concorde/Eagle/Intrepid, which is unnaturally roomy, and has striking stability and cornering. The same principle was then applied to the even bigger New Yorker and smaller Neon. And in addition to performance improvements, the cars get the bonus of the distinctive "cab-forward" look. (Like the Taurus aero look, the general cab-forward look was probably never patentable, although not exactly copyable. One has to wonder, however, if the cab-forward design concept, or aspects of it, were not at one time a candidate for some sort of utility patent.)

Even where this type of esthetic innovation cannot be patented or otherwise made proprietary, it can still be very good business. A good distinctive appearance can be quite effective for sales, while adding little or nothing to costs (especially to marginal costs).

This principle does not apply only to consumer goods. Even technical or industrial equipment sold to professionals to satisfy rigorous performance demands can enjoy improved sales from improved appearance.

This principle also applies to the user interface to software, which can benefit immensely from a good appearance, with an intuitive user-friendly look and feel.

Other strategies for patents are discussed elsewhere herein. (See Chapter 3 "Invent Around Your Competitor's Patent - And the Antidote - and Other Patent Strategies.") One of these strategies we call the toll gate strategy, which may be taken almost as a twelfth rule of invention-on-demand. However, the toll gate strategy may be the hardest rule to execute. It requires breadth of knowledge and deep synthesizing insight, and its pay-off can be a long time coming. The pay-off can, however, be large.

Recent Legal Developments

There are several recent legal developments in the field of "designing around" a competitor's patent that are important to note.

In a case decided March 31, 1993, *Westvaco Corporation v. International Paper Company*, 991 F.2d 735, 26 USPQ2d 1353 (Fed. Cir. 1993), rehearing denied. The court stated regarding successful IPC products for which IPC patented the structures embodied by the products: "Westvaco did not copy IPC's product, but instead attempted to design around IPC's product." The court went on to say that "designing or inventing around patents to make new inventions is encouraged. *See Wallace London v. Carson Pirie Scott and Company*, 946 F.2d 1534, 20 USPQ2d 1456 (Fed. Cir. 1991). Keeping track of a competitor's product and designing new and possibly better or cheaper functional equivalents is the stuff of which competition is made and it is supposed to benefit the consumer . . . it should not be discouraged by punitive damage awards except in cases where conduct is so obnoxious as clearly to call for them. . . . The District Court clearly erred in finding that Westvaco's infringement was willful because Westvaco reasonably relied on timely and competent opinions of its outside [patent] counsel and attempted in good faith to design around IPC's products. Accordingly, . . . enhanced damages may not be awarded." (emphasis added).

In this case, Westvaco monitored the patents of its competitor, IPC, and attempted to design around them. In doing so, it obtained a written opinion of non-infringement from outside patent counsel. In the end, the Federal Circuit found that Westvaco had taken the right approach but had, in the end, failed to adequately distinguish its products from IPC's products. However, Westvaco did avoid triple damages for willful infringement since it was under advice of outside counsel that its new products had avoided infringement.

There are several lessons from this case: (1) the "design around" strategy is encouraged and legitimate; (2) even when design-around products are "functional equivalents" in the market place of a competitor's patented product, the design-around product can avoid infringement of the competitor's patent; and (3) although the "design around" strategy is not guaranteed a success and may lead to infringement, the exposure for such infringement can be cut by two-thirds if the resulting new products are protected by a written opinion of outside patent counsel regarding non-infringement.

In *Read Corporation v. Portec, Inc.*, 970 F.2d 816, 23 USPQ2d 1426 (Fed. Cir. 1992), rehearing *en banc* denied, the court stated "In several meetings between Portec engineer Gerald Dahlinger and patent attorney Brett Valiquet in January and February, 1987. . . . [they] discussed several possibilities for designing around the claims of the '194 patent. . . . Of course, determining when a patented device has been "designed around" enough to avoid infringement is a difficult determination to make. One can not know for certain what changes are sufficient to avoid infringement until a judge or a jury has made that determination." In this case, Portec obtained a written opinion of its outside patent counsel that no infringement was constituted by the new products that designed around the patent. However, it turned out finally that this opinion was incorrect. However, the case goes on to say " . . . judgment with respect to the '194 patent does not make his advice regarding that patent incompetent. Read cannot point to any substantial evidence which indicates that Portec did not have a good faith belief that it was not infringing because it has successfully 'designed around' the '194 patent. Thus, the factors of willful infringement and copying are not present." Therefore, this case, indicates that it is good insurance in the "design around" strategy to obtain a written opinion of non-infringement from outside patent counsel in case the "design around" effort fails and infringement is actually constituted. This would, in effect, minimize the exposure for damages by two-thirds from what it may otherwise be.

In *Wallace London v. Carson and Pirie Scott and Co.*, 946 F.2d 1534, 20 USPQ2d 1456 (Fed. Cir. 1991) the court stated "Although designing around or inventing around patents to make new inventions is encouraged, piracy is not. Thus, where an infringer, instead of inventing around a patent by making a substantial change, merely makes an insubstantial change, essentially misappropriating or even "stealing" the patented invention, infringement may lie under the doctrine of equivalents." This language sets out the modern test for inventing around. Basically, the new invention must represent a substantial change over the patent to be avoided. If the new product only represents an insubstantial change, then it may constitute infringement under the doctrine of equivalents and lead to damages.

The leading modern case in this area is *Kimberly-Clark Corporation v. Johnson & Johnson*, 745 F.2d 1437, 223 USPQ 603 (Fed. Cir. 1984). This is one of the first cases of the then newly formed Federal Circuit. In this case, the court stated "We . . . agree that there is no infringement. . . . Defendants have successfully designed around [the] claims, as they had a right to do," citing the 1886 Supreme Court case, *White v. Dunbar*, 119 U.S. 47, 7 S. Ct. 72 (1886).

Opinion Letter of Non-Infringement

The cases cited above together show the proper path for inventing around, from a legal point of view. That is, attempt to make a substantial change rather than an insubstantial change, to achieve "inventing around" and avoid infringement. Obtain a written opinion from outside patent counsel that the new product does not infringe, to minimize the exposure for liability in the event that one's improvement is later judged to be insubstantial rather than substantial. Note, however, that the new product may be "functionally equivalent" in the marketplace in the eyes of the consumer, but still avoid the legal test for the doctrine of equivalents and avoid infringement.

3

Invent Around Your Competitor's Patent (And the Antidote), And Other Patent Strategies

"Necessity is the mother of invention."
—Poor Richard's Almanac

In the previous chapter, we have discussed how your company can invent-just-in-time, or invent-on-demand, using the Rules of Virtual Genius. (See Chapter 2 herein, "A Corporation Can Invent-Just-In-Time.") Of course, you would like to prevent your competitors from inventing around your patents. To do this, you also need the antidote to inventing around.

The Antidote to Inventing Around

There is an antidote to inventing around, to deter your competitors from inventing around your patents. This is the other side of inventing around your competitor's patents, that is, stopping the competitor from doing the same to you.

Inventing Around

The best way to do this is to invent around your own patent and then patent that, too, before your competitor can. Successfully done, this will block your competition from inventing around your patent by using the same approach as you found first.

This invent-around-yourself antidote can be done as part of your original patent application. That is, as a first iteration, the draft patent application for the original invention is written. Then the inventor, his patent attorney, and others on the team, can invent around this original level of invention. To do this they may adopt the role of a competitor and directly apply the Rules of Virtual Genius to invent around the first iteration, just as a competitor might. Then, when an invent-around solution is made, that can be added to the original application. (Or if the original application has already been filed, then the solution can be filed as a separate application, as a continuation-in-part, or a second generation improvement application.)

This process, if done after issuance of the first patent, can be viewed as an effort to develop second generation improvements before the competition does.

Why Is Inventing Around So Hard?

Why is "inventing around" or "designing around" so hard? Why is it not done more? Especially after reading the rules for inventing around (the "rules of virtual genius" discussed in this and the preceding chapter), the process seems straight forward enough. However, the process simply does not seem to actually get done that much, despite its obvious value. What is the bottleneck?

The answer seems to be the same as why any invention is the exception rather than the rule, that is, new ideas are simply rare. New ideas are hard to create. The rules discussed in this book for inventing, do not in themselves generate new ideas;

instead, they can only direct the process of generating new ideas into directions that will be the most valuable.

Generating new ideas can only be facilitated, but not caused, by knowledge (including, hopefully the analysis in this book) and intelligence. Generating new ideas seems to be simply an individual personality trait, a way of dealing with challenges, a turn of mind. All we can offer here is a way to direct the results of this process, to identify and evaluate the results that are valuable.

One lesson of this conclusion, is that when idea-prone individuals are available in or to an organization, they should be encouraged and their efforts commercialized as well as possible.

It is our impression that inventing around, despite its great value, is most often used merely as a legal argument to characterize certain situations, after the fact, as a part of a case for non-infringement. It appears to be far less done as a pre-meditated project including engineers and attorneys (partly because it is difficult), despite its great value. Inventing around should be done more as a planned project, rather that just as an afterthought legal argument.

Other Strategies

Inventing around your competitor's patent portfolio can be a very effective part of a response to a competitor's enforcement action against your company for patent infringement. Inventing around can also be a very effective method to prepare a new product introduction into a competitive market while avoiding liability for infringement. However, the response of designing around a competitor's patent is only one possible response strategy to a problem patent in your competitor's hands. There are several other strategies.

The Picket Fence Strategy

Another strategy is the picket fence strategy. The Japanese, by reputation, excel at this strategy. When your competitor has a key fundamental patent, you may invent a series of patents that represent small incremental innovations around the core technology. The incremental innovations represent the preferred products in which the core technology may be used commercially. They then become a barrier to the effective use of the technology by the owner of the original technology. The owner of the picket fence then is in a position to force a cross-license of patents to acquire the core technology for its own use.

This sort of picket fence invention strategy is relatively straight forward and can be facilitated by close contact between the patent team and the marketing and manufacturing divisions regarding consumer's perceived needs to commercialize the core technology.

The Toll Gate Strategy

A third possible invention strategy in response to a competitor's patent is the toll gate strategy. In this approach, the entire body of prior art, not just your competitor's, is reviewed and generally conceptionized to identify the direction in which it is developing. You then project the trend to anticipate future developments. Finally you leap frog the current developments to file the first patent application with very broad claims for the next generation of improvements, even when you may have only a vague concept of the best products to implement these improvements. This patent, when it issues, then can act as a toll gate to the industry when its actual products develop to that level of advancement.

The toll gate strategy may be taken almost as an extra rule of invention-on-demand, or invention-just-in-time. However, the toll gate strategy may be the hardest rule of virtual genius to execute. It requires breadth of knowledge and deep synthesizing insight, and its pay-off can be a long time coming. The pay-off can, however, be large.

The Submarine Strategy: Old and New

The toll gate strategy is sometimes combined with the submarine strategy. The traditional submarine strategy was changed by the 1995 amendments to the patent statute, but to understand the new submarine it is useful to understand the traditional submarine.

The traditional submarine strategy involved filing a broad patent application, and then keeping it pending with a string of amendments to the claims. (Pending U.S. applications are secret, like a submerged submarine.) The later claim amendments would then zero in on specific product developments by competitors that come later. (The claim amendments must be supported by the original broad disclosure of the patent specification. This process of later refinements to the claims has in the past been attacked under the label of "late claiming," but can now be allowed by the courts.) Once the market and the application have developed with specific products, the patent would be allowed to issue, or to surface like a submarine for all to see. Of course, a long application process could result by happenstance, and not by specific intent; an accidental submarine, as it were. The intentional submarine strategy was considered by some to be an inequitable abuse of the patent process, but strongly defended by others.

Amendments to the patent statute to implement the GATT agreements, which became effective on June 8, 1995, together with proposed amendments, change and weaken the submarine strategy, but do not eliminate it.

The effective amendments provide for patents filed after June 8, 1995 to expire 20 years from the date of application, instead of 17 years from the date of issue, among other changes. (No changes were made or are expected in the option for "late claiming." However, some case law may be developing, regarding the litigious and prolific patent applicant Lemelson, to the effect that extreme late claiming may violate the requirement of diligent prosecution and undermine the enforceability of any eventual patent.) The effective amendments then, for the first time, put a limit on how long a submarine application can stay submerged.

Proposed amendments include the publication of pending applications and permitting third party opposition to patents.

With the publication of applications, submarines would no longer be a complete surprise. However with the remaining "late claiming" option, new claims (that are supported by the original specification) could still appear, or "surface," at any time while the application is pending and before expiration.

We can say that the traditional "immortal submarine application" strategy has been replaced with a submarine of finite life. And when applications are published (assuming this amendment is made) the submarine application will be replaced by a new "submarine claim" that will surface within the boundaries of the original specification.

The new submarine claims will be a weaker offensive weapon than the old submarine applications, but the new submarine claim will still be a threat. That is, the new submarine will have a finite life of 20 years (instead of an indefinite life when measured from the date of application), and the specification will not be a surprise. But the submarine claims may be a surprise.

There will be new weapons against the new submarine claims. The first weapon will be regular monitoring of the publication of patent applications, to detect dangerous new patent applications. We can also expect development at some point of ways for third parties to contribute to the examination of pending published patent applications, although it is unclear how they will develop. The U.S. Patent and Trademark Office does support an approach using third party expanded re-examination.

How to Submarine a Picket Fence

Kimberly-Clark can be read with another leading case of that year that facilitated the submarine strategy and refinement of claims, where supported, to cover new products, that is, *Railroad Dynamics v. Stucki Co.*, 727 F.2d 1506, 220 USPQ2d 929, (Fed. Cir. 1984). The court refers to "the inappropriate and long ago discredited late claiming [attack]," saying "the sole question . . . is whether the claims entered by amendment were supported by the disclosure in . . . . [the] original application." *Kimberly-Clark* and *Railroad Dynamics* together support an invent-around submarine strategy, consisting of a broad invent-around application, followed by subsequent product oriented fine-tuning, where supported by the original claims. This approach to alert patent prosecution can also serve as an antidote to a competitor's picket fence response strategy to your own basic patent. That is, you can submarine the picket fence.

The Counter-Attack Strategy

Another response strategy to a competitor's problem patent that can be pursued simultaneously is to attack the problem patent itself. Call this the counter-attack strategy.

A surprising number of issued U.S. patents may be improperly issued and can be subject to efforts to cancel or restrict them. This is despite the fact that there is a statutory presumption of validity for issued U.S. patents.

The first step of this strategy is to find a potentially fatal weakness in your competitor's patent by studying the prior art to determine what practices proceeded the patent. Another research step is to study the patent application file to determine if there is file wrapper estoppel, or what may be called "lexicographer's estoppel" in the file. If this is found, then the legal effect of the patent may be narrower than the patent appears on its face. (In a sense, the printed patent is incomplete in that the contents of the publicly available patent application file may make the legal effect of the printed patent narrower than the published patent appears on its face.)

Where the competitor's patent is found to be weak, a counter attack may be launched. If you are in reasonable fear of immediate litigation alleging patent infringement, then you may have grounds to file a motion for a declaratory judgment in federal district court, seeking a ruling that your competitor's patent is invalid.

The Stealth Counter-Attack

A cheaper and perhaps faster method than action in federal court, where it is available, is to file a request for re-examination in the U.S. Patent Office requesting that the competitor's patent be limited or cancelled. To do this, a request for re-examination should be filed with the Patent Office citing published references that create a new issue of patentability. The PTO filing fee in these matters is $2,200. The Patent Office then has three months to determine if a new question of patentability is present. If such a question is present, the patent owner has two months to respond to the request for re-examination. The requestor then has two months to respond to the patent holder's response. At that point, the examination of the application is re-opened and proceeds ex parte between the examiner and the patent holder, although the requestor receives copies of all the Patent Office actions. The re-examination may result in the patent being cancelled or the claims to the patent being restricted by amendments.

It is important to note that the request for re-examination may be filed by a patent attorney without revealing the name of his client, who wishes to remain confidential. Therefore, not only can the request for re-examination be cheaper and faster than litigation in federal court, but it also keeps the identity of the requesting party secret. You might think of this as a stealth attack on the life of the competitor's patent. That is, the competitor is aware that his patent is being attacked, but he cannot determine who the active party is. Therefore, the business relationship of requesting party with the patent holder need not be undermined by the re-examination.

It is also important to note, that on some points it is also easier to make progress against a patent on re-examination, than in federal court. The patent statute gives a presumption of validity to an issued U.S. patent, and that is the standard used by federal courts in a validity attack. However, on a re-exam, no such presumption applies. That is, on re-exam, the patent owner must prove that he deserves a patent, but in federal court, the threatened infringer would have to prove that the patent owner does not deserve a patent, if the patent is to be cancelled. Also, a re-exam can make progress by amending a patent to limit it, whereas the federal action can only strike a claim, not amend it. Also, if the re-exam fails, further attacks, usually on other evidence, can still be pursued in federal court.

The Cut Your Exposure Strategy

Another approach that may be used simultaneously with a problem patent is to obtain for your own product, where possible, an outside patent attorney's written opinion of non-infringement of the competitor's patent. Such a written opinion may reduce the liability exposure of an infringer by two-thirds, in those instances where patent infringement liability is subsequently found.

The Bargaining Chip Strategy

Another simultaneous strategy that may be pursued against a troublesome patent of a competitor involves developing bargaining chips in the form of your company's own patent portfolio that may be cross-licensed or traded with the competitor for a license in his patent. One way to initiate this program is to execute "an intellectual property audit" which reviews the existing and planned products and technologies of your company to ascertain what might be patented to create future problems for the competitor.

A troublesome patent by a competitor may be learned of in several ways. One way is that you may be monitoring the patent activity in your industry, to get immediate warnings of what patents a competitor has received. You may also monitor publications of foreign patent applications by your competitor, on the theory that he has made the same application in the U.S. Another way to learn of a problem patent may be by receiving a "nice" letter from your competitor informing you of the patent and the opportunity to obtain a license at a royalty rate to the competitor, as an alternative to his being "forced to disrupt the industry." A third and more unpleasant way in which you may learn of the competitor's patent, is to receive a "cease and desist" letter which alleges that your products are infringing another's patent and demanding that the infringement stop immediately. This particular type of letter, unlike the "nice" letter discussed above, may place you in reasonable fear of immediate litigation regarding infringement. This, in turn, allows you file your own action for declaratory judgment in federal district court to have the patent declared invalid, in those circumstances where you may feel this is the preferred path to pursue. It is important to note that where a competitor is taking an aggressive role in attempting to force his patents against you, it is possible for you to take aggressive action against the competitor by attacking the validity of his patent, and by developing a patent portfolio that will give you the weapons to take your own aggressive posture in the near future. Of course, the immediate letters of the competitor must be responded to in a defensive manner, but offensive action may be simultaneously pursued.

4

Prevent Product Re-Use With Patents

"For Single Use Only. U.S. Patent No. 1,234,567."

Patents can be used to enforce disposability and non-reuse of disposable products.

It is hard to imagine how non-disposability *per se* can be patented, but a creative use of patents and labels on products can make unauthorized re-use of a "disposable" product constitute patent infringement, and give rise to damages and injunctions.

An economic problem for some disposables is that they are often repeatedly re-used, rather that repeatedly repurchased after only one use. The re-use may also raise safety and liability concerns, in addition to the obvious economic problem for the manufacturers. Fortunately, three new patent law cases indicate how to solve these problems with techniques of patent law, in the case of patented disposables or would-be disposables. With careful patent drafting and product labeling, it is now possible to use patent law to require and enforce single use of patent-related disposables.

Indeed, these patent law techniques can "invent" and enforce disposability for patented devices, where objective engineering analysis would see no inherent disposability. Furthermore, these patent techniques are so powerful that they may be able to enforce other unrelated limitations on the use of patented products, whether or not they are disposable.

The Label Case

In *Mallinckrodt v. Medipart*, 976 F.2d 700 (Fed. Cir. 1992), the court held that "single use only" labels on patented disposable medical products can now be enforced by U.S. patent law. Unauthorized re-use of these products can constitute patent infringement. The reasoning of the case also indicates that other "label restrictions" on patented products, besides "single use only," can be enforced under U.S. patent law.

In the events leading to this law suit, Mallinckrodt manufactured a medical device for hospitals called an "Ultra Vent." Parts of the device were patented and intended for one use only. These parts were marked with the applicable patent numbers, and the inscription "single use only." The package for the disposable provided that each unit was "for single patient use only." We can assume that actual single use in the market place was important to Mallinckrodt for economic, safety, and liability reasons. Re-use would cut Mallinckrodt's sales, and might adulterate the product to cause injury to patients and unjust liability exposure to Mallinckrodt.

However, certain hospitals which purchased the products from Mallinckrodt were not disposing of the disposables, but instead they were shipping the disposable assemblies after use to Medipart, Inc. Medipart cleaned and repackaged the disposables and sent them to radiation sterilizers for sterilization. The "reconditioned" units were then shipped back to the hospitals from which they came for further use. The "re-conditioned" assemblies continued to bear the description "single use only" and the trademarks "Mallinckrodt" and "Ultra Vent."

The court found that when an authorized manufacturer sells a licensed product, they sell the product with an implied license for the buyer to use it and to repair it, but not to reconstruct it. On the implied license theory, the court concluded that if a use license is implied, then that the implied license can be explicitly limited by a writing. The court found that the writing on the product itself, "single use only," was an explicit written limitation on the implied use license, and was enforceable. Therefore, any use by any party beyond "single use only" was outside the patent license, which was implied in general but specifically limited in writing, and such additional use constituted patent infringement. Therefore, the manufacturer Mallinckrodt could act against any party participating in the multiple use, including Medipart, the hospital, and the sterilizer. The cause of action would be patent infringement, and the remedy could be a permanent injunction to stop the re-use and damages (which could be tripled in some cases).

The court did comment that as in any license arrangement, certain antitrust issues could be generated by specific license provisions. An example of these questionable provisions would be tie-in agreements. (For example, a label like "for use with XYZ Co. parts only" might in some cases raise tie-in type anti-trust questions.) However, there were no tie-agreements in the subject license and the label was upheld.

An extension of reasoning of this case would indicate that any provision that is acceptable in a written license agreement, could be incorporated into a patent-enforceable label on a patented device, besides simply the "single use only" label. For example, labels may be able to restrict devices, whether disposable or not, to use in certain geographic territories or use for certain limited time periods. (A label limitation of a product to be used only in conjunction with other products by the same manufacturer may raise, in specific instances, anti-trust questions connected with the tie-in license doctrine. However, this concern might be mitigated in a context where such use limiting other equipment was necessary for medical safety and health reasons.)

No Need to Sue a Client

Note that in the *Mallinckrodt* case there was no need for the patent holder to sue a client. Instead, the patent holder sued a competitor servicing the patent holder's client (as is found in almost all patent infringement cases). Mallinckrodt did not sue a hospital that bought its equipment. Instead, Mallinckrodt sued Medipart, which was a competitor in the hospital market.

A Tool Against the FDA?

These patent law developments are interesting in cases where the FDA encourages manufacturers of disposables to provide information to facilitate the reuse of disposables against the wishes of the manufacturer. Under the *Mallinckrodt v. Medipart* doctrine, instructions for reuse pursuant to FDA requirements may not constitute an implied license for reuse, where the product itself carries "single use only" labels. In other words, even though the FDA may require instructions regarding cleaning and reuse, this would not require the manufacturers to actually grant the license for such reuse or refrain from taking action under patent law for patent infringement for such reuse. Indeed, the FDA may exceed its authority if it were to require a waiver of patent rights by any sort of requirement for a compulsory license to reuse a patented product.

The Little Piece Case

In a second case, an important point is made regarding the patenting of disposable medical products. See *Surgical Laser Technology v. Surgical Laser Products*, 25 USPQ2d 1806 (D.C. Penn. 1992). This indicates, in the case of disposables, an important technical distinction that must be followed where possible, in the drafting of patent claims. The case makes clear that a disposable component of a larger patented device may be copied and/or reused without infringement, where no independent claims are drafted strictly to the disposable component. That is, wherever possible within a patent, a separate independent claim or patent should be drafted to the disposable component of the product. If the independent claim is drafted to the entire device, including disposable and non-disposable elements, then the manufacture of the disposable by another party would not in itself constitute patent infringement, because it would avoid the reusable portions of the independent claim. Where an independent claim is drafted only to the disposable components, then anybody manufacturing the disposables in competition with the patent owner would infringe that independent claim.

(Note, however, that selling a disposable component of a patented machine, can constitute contributory infringement, if the component is a material part, especially made for such use, and is not a staple article in commerce suitable for non-infringing use. See 35 U.S.C. 271.)

Virtual Disposability

Combining the teaching of *Surgical Laser* and *Mallinckrodt*, one understands that the patent law can be extended to make "single use" or in that sense "disposable" any product for which a patent is obtained. To do this, the "single use only" label should be attached to the components described in any independent claim so that any reuse or competing manufacturer by them would constitute infringement. This would be true even where the engineering characteristics of the product would actually permit re-use. Call this legal disposability, or virtual disposability.

The corollary to the *Surgical Laser* case is that any product for which there is an independent claim could be made "disposable" in the legal sense if a "single use only" label were attached to it.

The main legal upshot of making reuse of a patented product constitute patent infringement would be that the patent law could be used to obtain injunctions against further reuse and to obtain cash damages for past reuse. Note that these actions can be taken against any party that makes, uses or sells the infringing re-used product.

A Further Extension of Patents for Disposability

A surprising 1994 medical device patent case extends these concepts for the use of patents to enforce disposability and to prevent competition in making replacements for disposable medical parts (even where the disposable parts are not themselves patented).

In *Lifescan v. Can-Am Care*, 31 USPQ2d 1533, the use of license restriction labels on packaging regarding patented products and methods is further elaborated. In this case, Lifescan made a blood glucose meter for home use, which required that the patient insert a one-use strip in the meter. Lifescan manufactured the meter and the replacement disposable strip. Lifescan obtained two patents, nos. 4,935,346 and 5,049,487, for methods of using the meters, but not for the strips themselves. After selling the meters for awhile, Lifescan began marking the meters with the patent number and with the following label. "Purchase of their device does not give a license to practice these patents. Such a license is automatically granted when a device is used with the enclosed or separately purchased [Lifescan] disposable test strips. No other test strip supplier is authorized to grant such a license." The strip itself was not patented. Can-Am began manufacturing and selling the strips for use in the meter with labels which said in part "Due to the patent infringement issues alleged by Lifescan, the test strips in the box should not be used in those [Lifescan] meters which contained the [Lifescan] notice."

The general rule is that the authorized sale of a patented product carries with it an implied license to use and reuse the product; however, labels on the product may restrict that implied license. In this case Lifescan claimed that its restrictive label extended to the use of the unpatented strip in the patented device. Furthermore, Lifescan claimed that the unpatented strip sold by Can-Am induced the infringement by the ultimate user, and hence Can-Am was guilty of "inducing infringement (contributory infringement) and unfair competition by actively and knowingly aiding and abetting other's direct infringement."

Can-Am tried to avoid trial by a motion for summary judgment. This motion was denied and the case was directed to go to trial. The judge, in ruling on the motion for summary judgment, found that the original Lifescan label limitations may be valid, and that the warning labels by Can-Am may in fact be inadequate to avoid induced infringement by Can-Am. Can-Am can be expected to argue that the license limitation was over-reaching and tried to extend the patent to devices not covered by the patent, that is to the strip, and that this constituted unfair competition and patent abuse. However, Lifescan can be expected to argue that any such extension might be proper under law and a matter of good public policy since it encourages the safe and effective use of the medical device and does not endanger the patient with disposable strips that may be defective. Furthermore, Lifescan would be pursuing its legitimate interests in avoiding exposure to vicarious liability for injuries caused by defective Can-Am strips.

Any further development of this case could be very important for the protection of disposable medical products that are not directly protected by independent claims in a patent, but are used in a broader system that is protected by patents, where the proper label limitations are used on the product. Hence, for example, a larger system (such as a monitoring system for disposable bio-sensor catheters, or an artificial blood pump with disposable blood bags and tubes, where the patents cover the overall system but not the disposable component itself), the system patent with proper labelling could prohibit competitors from manufacturing competing disposables to be used in the overall unit. This case bears watching and it can be very important for the disposable medical industry.

Note also that in this case there was no need for a patent holder to sue a client. Instead, the patent holder sued a competitor. Lifescan did not need to sue a buyer of its meters or strips. Instead, Lifescan sued Can-Am, a competitor in the replacement strip market.

Restricting Anything, Not Just Re-use of Devices

Note that these enforceable label restriction may be added to by any patented technology or device to restrict its use. Hence these label restrictions may also be used with, for example, software and pharmaceutical drugs, in addition to devices. Furthermore, these enforceable label limitations might be used to limit the use of the product in any manner that would be acceptable in a written license agreement. Hence these label limitations might extend, for example, to limitations as to the industry, the geographic area, or the purposes for which the labeled product can be used.

These new developments are a breathtaking new way to control the aftermarket and economics of patented products, and its impact will eventually be felt in many industries.

5

Get Your Patents on the Fast Track

"Time is money."

A procedure that can reduce the time required to process a new patent application is available but not widely known.

Within the patent office, there is a fast track and a slow track, and unless you request otherwise, you get put on the slow track.

To get on the fast track, an inventor must petition for and be granted an accelerated examination. This process moves the application to the front of the line, but limits the interaction between the applicant and the examiner to one amendment cycle unless further fees are paid to the patent office.

A petition for an accelerated examination must be customized for each application and include the results and analysis of a patent search.

Although legal costs for this process can be higher, getting a patent faster can be important in obtaining venture capital and bringing a product to market. That is, if you have a person with a checkbook waiting to give you money, it may be worth a few thousand dollars to get rid of a two-year wait.

This expedited procedure may also become more important, since June 8, 1995 has passed. For patent applications filed after that date, the term of any eventual patent is 20 years from the application date, instead of 17 years from the issue date. Hence, the longer the patent application is pending, the less time the actual patent will last. This will encourage use of the expedited procedure in some cases to increase the term of enforceability of the patent by getting the application through the Patent Office sooner. (Patents are enforceable only between their issue date and their expiration date. Patent applications, the so-called "patents pending," are not enforceable.)

There is also some thought that an accelerated patent application may have a better chance of easy passage through the Patent Office.

While inventors can apply for acceleration with any patent application (if the inventor does a patentability search and analysis of the prior art), the patent law favors accelerating inventions relating to environmental quality, energy conservation, safe research of DNA, and superconductivity. An inventor's poor health or advanced age, manufacture contingent upon a patent, and infringement also can accelerate a patent application.

This expedited procedure is not the best procedure for all applications. Indeed, there can be as many reasons (and as many ways) to delay an application in the Patent and Trademark Office as to expedite it. However, in some cases the expedited approach can be the perfect tool for the job.

6

Eight Tips for Patent Licenses

"Patent what you can sell. Sell what you can patent."

There are a variety of books and articles available which survey the law of negotiating and drafting licenses for patents and other transactions regarding technology. However, in the process of executing a variety of these projects over the years, we have found a trend of specific weaknesses in many negotiated contracts and agreements that can present particular problems in the future. Here we offer comments on these.

Write Patents That Are Worth Paying For

Patents should be written and prosecuted before the Patent Office, with an eye to giving them the maximum possible value in the market place, whether for licensing or otherwise. The point of applying for a patent is not just to get a patent, but to get a patent of value that can earn its way in some manner after issuance. There are several rules of thumb in doing this.

Write the entire patent, especially the claims, in plain English, as much as possible. If the patent will ever be licensed, then normal people will have to read it and decide to write checks. If the patent is ever enforced against an infringer, then a judge (probably with a B.A. in literature), and maybe more normal people on a jury, will have to read it and decide that it means something. If technical jargon must be used, try to make it common jargon of the field of the invention, not patent legalese.

Some fundamental patents are of great value, but most patents of value claim specific recognizable products, at least in addition to broad claims. The value of patents is to protect markets, and products are what is sold in markets, not big ideas.

Where possible, claim components that are sold separately. This can facilitate licensing of specific components to separate licensees, and make enforcement against component infringers easier.

Include software claims when possible, even in products that are primarily not software products.

In claiming pure software, use several different forms of software claims. Claim the software on a floppy disk, in addition to other forms of claims.

Consider claiming the computerized execution of any method.

After the patent application, and the claims are laid down, have the inventor go back and try to invent around his own invention. Take these new results and include them in new claims in the application. Then file the application. (If the competition is good, someday they may try to invent around your patent. It is better if you get there first and pre-empt the invent-around result, for your own portfolio.)

Then after you think you are finished, step back and write some broader claims.

Price, Formulae, and Other Mathematics

It seems that people in general, and attorneys in particular, can be vague and imprecise in clarifying quantitative ideas and expressly them clearly in English. We have found that it is often the case that price terms in a license or other negotiation, which is inherently arithmetic, can be vague or ambiguous. This is particularly the case when price is determined not as a single fixed amount per a unit of product, but is the result of some sort of formula.

There is no magic answer that will address the situation in all cases regarding this problem. We must bear in mind to give special attention to all quantitative and arithmetic expressions in contracts. The ultimate test in reading these documents is to determine if they precisely and unambiguously describe in all circumstances the particular number and the process by which it is to be computed that is to be placed on a particular check and made payable to the order of a particular party. The ultimate test of meaning and communication for this language is that when a party goes to write a particular check in question to another party pursuant to the contract, that the contract gives that party no flexibility in determining what that number is. This is the single acid test to which all arithmetic concepts and descriptions in a price term should be addressed.

Terms of Payment

In technology transactions it seems that the timing payment is often left to be somewhat flexible or ambiguous. Although this may not be foremost in the minds of the negotiators on the front-end of a transaction, it can be critical or even life-threatening to one or both parties to a business transaction in the execution of a contract after the negotiation. Indeed, timing of payment questions can make or break the profitability of a negotiated transaction. Regarding, for example, royalty payments pursuant to a license, are the payments for a given year made at the beginning of the year or at the end of the year? Are they made monthly, quarterly, or annually?

Regarding the manner of payment, particularly in international transactions, is payment made in U.S. dollars by a check on a U.S. account of a U.S. bank? Are the funds wire transferred in dollars? Is any check used, certified or cashier's check?

Are payments made subject to any offsets, discounts, returns, change orders, or other factors that may affect the amount of payment?

What constitutes late payment and what is the financial effect of a late payment?

Many recipients of payments believe that they have been paid when funds are in the recipient's accounts and are payable by the recipient to other parties. Many payors would prefer that the funds are considered paid when a check or other order for payment is issued by the payor. However, for example, when an international payment is made, the difference between these two definitions of payment may routinely be six weeks in some circumstances.

What Products Are Covered by the License?

The difficult aspect of this question is that parties usually negotiate transactions in contemplation of their current products. However, the term of a license or other agreements may be quite long and involve technology that is dynamic and has a relatively short product cycle. Perhaps software is the most extreme example of this sort of product. This highly malleable and accelerated rate of evolution for software products is particularly important as software becomes the nexus of innovation in more and more products and smart equipment. How is the product defined, and is the definition of product proper to include the intent of the parties regarding future variations of the product? When will the product have evolved incrementally to the extent that it is a distinct product and outside of the scope of the contract? What about jointly developed modifications or modifications requested by one party to be made by the other party? For example, licensees often request licensors to develop specific variations of the product.

There is no easy answer to this question, and as is the case with many of the other questions in this chapter, perhaps the ultimate answer is simply that they are ongoing questions for the management and control of a contractual relationship, and cannot be necessarily fully and completely addressed and disposed of once and for all in the front-end of a contractual relationship.

Ownership

It is important for a license or other contractual arrangement between parties regarding technology to specifically describe what ownership rights are transferred, if any, in patent, copyright, trademark, and other intellectual property related to the products, as opposed to ownership of specific unit embodiments of the product. For example, regarding the purchase or license of program smart equipment, does the licensee merely own the specific units of hardware delivered? Were they licensed to use the embedded software, or does he also own copyrights to the software? This may affect the ability of the "buyer" to resell the product with the necessary software, or the modified software, for internal purposes, or modify and sell the software to others.

The ownership question is probably the most difficult and important for variations as products develop after the original signing of the license or contract. Questions that arise are: Who develops the subsequent improvements or variations? Who is to own them? What if they are developed strictly by a licensor, or strictly by the licensee, or jointly?

Bear in mind that technology is like plant or animal life: it evolves almost as if by a separate force of nature. The difference is that many technologies, particularly software, or software/hardware hybrid technologies, evolve much more rapidly than natural biological life. It is often mentioned that the average product life of an electronic hardware product is now about 18 months, and we can assume that many software product lives are even shorter between one release and the next.

The licensor, of course, wants to retain title to all variations, modifications and additions, whether conceived of by the licensor, the licensee, or the two parties jointly, and to resolve all questions regarding the definition and extent of the controlled products in favor of ownership by the licensor. The licensee may have much the opposite orientation, being particularly interested in owning or sharing developments or variations that the licensee specifically thought of and requested, and in many cases paid for. These issues are usually resolved more by a balance of market power than by any inherent concept of fairness or equity, in those cases where they are explicitly negotiated and written into the contract.

An interesting quirk in these ownership questions is often found in the case of financing programs involved with federal funding. In programs involving federal government funding, it is often possible for a private corporation funded by special federal programs to, in the end, obtain almost exclusive ownership of subsequent developments that are the results of the federal funding.

We have found that ownership of intellectual property funded by federal programs is seldom offered by the federal funders to participating private companies, and must be requested and negotiated for by the private companies receiving the funding. However, we have been successful in obtaining ownership of later developed intellectual property for the private companies, where such ownership has been sought and pressed for.

The Exclusive License

An "exclusive license" may also exclude the licensor. Many people seem to think, wrongly, that "exclusive" here just means that there is only one licensee, and that one licensee can practice the licensed art with the licensor. To the contrary, "exclusive" probably means, in the absence of language in the license to the contrary, that the licensee is the only party that can practice the licensed art, to the exclusion of all other parties, including the licensor. This can be a nasty surprise to an unsuspecting licensor. Language to the contrary in the license can take care of this issue in advance signing the license.

The Non-Exclusive License

The non-exclusive license may be personal, unless it states otherwise. That is, the non-exclusive licensee may not be able to assign or sub-license its rights, unless the license document explicitly grants this right.

Taxes

Important tax planning issues can be presented and disposed of by licenses, even if these issues are not explicitly dealt with by the parties.

The taxation of patent licenses is an important but complicated subject that will not be treated in detail here. Tax law for patents is also quite dynamic, and several proposals are currently before Congress that would change important aspects of the taxation of patent transfers.

The main tax issues for patent licenses arise from the fact that there are a variety of options for structuring these transactions. The choice of a specific structure can have large impacts on the tax obligations generated for each party.

This is a difficult subject to generalize, but we can say that the biggest tax structuring question tends to be whether the transaction should be structured as a capital gains transaction or an ordinary income transaction. This can be a point of hard negotiation, since the "buyer" and the "seller" in the deal tend to want opposite approaches for their own tax purposes.

Basically, if substantially all of the patent rights are transferred in an assignment or license (and if other conditions are met), then both parties may treat the transaction as a capital gain transaction. Otherwise, the transaction tends to be an ordinary income deal for both parties.

In a patent transfer transaction, the "seller" tends to prefer capital gains treatment. This allows royalties received to be offset by any basis, and then the gain may be treated as long-term capital gains. (Currently, individuals have lower tax rates for long-term capital gains than do corporations and individuals pay lower tax rates for long-term capital gains than for ordinary income. In the near future, corporations may get similar benefits.) In an ordinary income transaction, royalty receipts are taxed at ordinary income rates.

On the contrary, the "buyer" tends to prefer ordinary income treatment for patent transfers. In such a transaction, the buyer may expense its royalty payments (instead of capitalize and amortize them, as they would with a capital transaction).

Among other things, these tax considerations tend to make the transferor of patent rights favor an assignment or exclusive license, with a separate license-back transaction of any limited rights; as opposed to the option of a single limited license with rights reserved to the licensor. The first transaction may allow the transferor to get capital gains rates for royalties received for the assignment, and ordinary income expense deductions for royalties paid for the license-back. The licensee, of course, may resist this structure because it may get the reverse treatment. That is, the licensee may see capital gains treatment for the royalties paid for the assignment (requiring the licensee to capitalize and amortize the amounts, instead of using current expense treatment), but ordinary income tax rates for royalties received for the license back.

Other tax issues that may be of interest may include timing of tax realization (and installment sale, and open transaction questions), possible imputed interest, disallowance of losses between related parties, and collapsible corporation issues.

Tax questions for intellectual property other than patents, including copyrights, trademarks and trade secrets, can be equally important and complex. Transactions in these properties present many of the same general tax issues, but they can follow somewhat different rules regarding specific issues.

Licenses involving one or more foreign parties raise a variety of distinct questions regarding U.S. taxes for foreigners and foreign transactions. Also, of course, foreign tax questions are raised. Here, as with strictly domestic transactions, the benefit of an investment in tax planning in structuring the transaction can be large.

7

Patent Due Diligence for the Finance of Technology Companies

"My best ideas are somebody else's."
—Benjamin Franklin

Due diligence and full material disclosure for patents and other intellectual property has not, traditionally, been aggressively pursued. However, the increased value of patents initiated by the amendments to the patent statutes in 1983, and as shown by recent patent litigation, has made it necessary to pursue patent due diligence aggressively. (See, for example, the recent "bet the farm" patent litigation in *Polaroid Corporation v. Eastman Kodak Company*, 867 F.2d 1415 (Fed. Cir. 1989) and *Stac Electronics v. Microsoft Corp.*, D.C. C.Cal. CV-93-413-ER, May 13, 1994, June 8, 1994.)

A key to patent due diligence is that every problem has a solution, and the sooner the solution is found, the cheaper and easier it usually is. The point is to find solutions, not just problems.

For a technology company, the goal is ongoing activity to develop and maintain a patent position of financiable quality, that will pass due diligence. The immediate impact of this is to increase sales volume, increase profit margins, increase net income, legitimately suppress competition, and achieve other business plan targets. The ultimate effect of this is to increase the net value of the company.

Many companies are financed that have major assets in intellectual property ("IP") owned as patents, copyrights, trademarks, trade secrets or confidential information. Such financing includes public and private issuances of securities, secured lending for which IP assets are taken as collateral, mergers and acquisitions, licensing, and joint ventures. In these financial transactions with IP, basic requirements of due diligence and full material disclosure must be properly handled regarding ownership (also called "title"), validity, scope, infringement and other questions about patents and other intellectual property. There are specific steps available by which IP assets may be evaluated. This is essential in any financial transaction in a technological industry, where due diligence and full material disclosure are required.

Traditionally, there has been poor communication between patent attorneys and other attorneys involved in securities, corporate and lending practices. Partly this is because of the past tendency of patent attorneys to practice within patent specialty firms. This limited their practice to obtaining patents from the U.S. Patent Office and litigating infringement issues regarding those patents. However, today diversified full practice law firms with large securities, corporate and lending practices have begun, in some instances, to develop intellectual property departments. With the new trend, some patent attorneys within these practices have developed experience and expertise in financial transactions involving technology companies. This presents an opportunity to correct old practices regarding title to patents and related evaluation issues.

It is sometimes surprising to securities and corporate attorneys, or to attorneys who have dealt with commercial real estate situations, to experience the quicksilver nature of ownership to intellectual property (patents, trademarks, copyrights, trade secrets and confidential information). This is particularly surprising in comparison with the complete statutory title regimes for real property and personal property. Title for intellectual property can be examined, to a degree, and title problems can often be identified and cured. (And if not cured, at least pointed out for investors to make their own risk assessment and adjustment in their offering price).

Fourteen Steps for Patent Due Diligence

The fourteen steps discussed below are available to deal with patent due diligence. Not all the steps are applicable, or affordable, in every case (or perhaps even in most cases), but they represent the basic menu of choices.

Step 1. Title Search. Basic steps in a title search for intellectual property include review of (1) all written contractual agreements with employees, consultants, suppliers, and other third parties, (2) all non-disclosure agreements with parties with which negotiations took place and to whom proprietary information was disclosed, and (3) the so-called "assignment records" of the U.S. Patent Office. The main question of this search is to determine if the affirmative steps were taken to create a written recorded paper trail so that title for technical developments rests in the corporation rather than the individual employee or consultant inventor. If this is not the case, the title of record can be completed by obtaining and recording written assignments for all appropriate rights. This is analogous to a title search in real property or personal property.

Step 2. Title Opinion. Patent title opinions can be given by a competent independent patent attorney, in much the same way as title opinions or title commitments are given for real estate. In the patent title opinion, the specific steps taken to review title and the specific documents reviewed should be itemized and the opinion limited to the same. The specific problems found with title, if any, can be indicated and excepted from. The final title opinion can be given subject to the indicated data base and exceptions, with the proper caveats regarding lack of assurances regarding any future litigation of the same questions, the operation of laws such as bankruptcy, limitations on parties and purposes for reliance, and otherwise. This is analogous to a real estate title insurance binder.

Step 3. Validity and Patentability Opinions. Another question related to patent title that can be analyzed by independent patent counsel is the question of patent validity. Pursuant to Section 35 U.S.C. 282, patents issued by the U.S. Patent Office are presumed to be valid. However, many patents are issued that are invalid and they can be so declared by a federal court. Also, it is common in the case of an infringement lawsuit for the defendant to cause the Patent and Trademark Office to re-examine the patent application (which may result in the patent being cancelled). Competent outside patent counsel can review the patent application files and prior art for important patents and develop written opinions regarding the validity of issued patents, pointing out the problems and main attacks on validity, if any, to facilitate the evaluation of the patents. An analogous activity for pending patents is to develop an opinion of patentability regarding a pending application. The items in this step are also analogous to a real estate title binder.

Step 4. Scope of Patents. Another question that patent counsel can undertake to evaluate patents in a financial transaction would be to analyze whether the products and services of the corporation are actually protected by its patents and patent applications pending. That is, does the patent portfolio of the company actually work to suppress competition and protect the lines of business of the corporation. A company's products may simply be outside of the scope of its patents. Again this is the sort of opinion that is rarely given by patent counsel in a financial context, but can be of critical importance and necessary for due diligence and full material disclosure. Opinions can be issued on this question.

Step 5. Infringement Opinions. Another question that should be analyzed in financing a company in a technological industry is "Do the products and services of a corporation infringe on the patents of others?" Where a product or service is being provided, even when it is within the scope of the patents owned by the providing company, it may still infringe the patents of others. Patent infringement can result in an injunction to stop doing business, a judgment for damages, and a tripling of damages. A written opinion of non-infringement by outside patent counsel, even when wrong, can serve to avoid a tripling of damages where liability for infringement is eventually found. This opinion letter process can also be a prophylactic step to cause the company to take steps to avoid infringement before it occurs. Such opinion letters also may act to limit personal liability of the officers and directors for corporate infringement. Steps 4 and 5 are analogous to a real estate boundary survey to see if the property improvements encroach on the land of others. Step 5 is also analogous to an insurance policy against infringement damages, with a 33% deductible.

Step 6. Is this Patent Too Narrow? Another question that should be addressed is "Can the patents owned by the company be bypassed by others or invented around without infringement?" In other words, can the market niche and customer base of the company be accessed by competing products and services that do not infringe upon the patents in question? This is related to the question of whether or not the existing patents of the company succeed in protecting the products of the company and suppressing competition. The "invent-around-it" strategy is often pursued by competitors in response to a problem patent in the hands of others.

Step 7. Disclosure. The patent portfolio owner should also be made to disclose any IP litigation that has been threatened, and to deliver any prior opinion letters that the owner has received.

Step 8. Engineering Tests. Particularly for new products, and most of all for software, independent engineers should test and report on the functionality of the product. Mock-ups of incomplete software can be very difficult to spot in any controlled "demonstration."

Step 9. Liens. Another question that may be addressed by the patent title opinion, where applicable, concerns the validity of secured liens and title acquired through foreclosure of security interests.

A little background on the patent statute is important here. 35 U.S.C. 261 (the title section of the U.S. patent statute) states that patents are to have the attributes of personal property except as otherwise indicated in the statute. However, the same section goes on to state that title to patents arises with the inventor and remains with the inventor until it is transferred by written assignment. Furthermore, the statute provides that assignments may be recorded in the public records of the assignment branch of the Patent Office and that such recording services as notice to subsequent purchasers, if the assignment is filed within three months of its date. No other instruments of title are explicitly addressed in the statute. However, in the regulations promulgated by the Patent and Trademark Office under this section, it is stated that all instruments of title will be accepted and filed in the "assignment" records, including licenses, mortgages, and liens. However, this same regulation states that recording of instruments of title other than assignments may not serve as constructive notice to the public because they are not enumerated in the statute itself. Furthermore, the statute says nothing about the perfection of security interests, the procedures to foreclose and liquidate security interests in patent collateral, questions of deficiencies, or rights of redemption. There is a body of cases that attempts to address these issues, however, the bottom line is that the security and specificity of a complete statutory title regime is simply not present and cannot be provided by the inchoate case law approach. Common practice is not to record patent licenses anywhere. The result of all this is that it is impossible to do a complete or reliable title search for patents, because of the lack of a complete statutory title regime, as we have with real property and personal property.

Given this statutory mess, the most secure way to perfect a security interest in patents currently is to simultaneously do two things: (1) file a document in the "assignment" records of the Patent Office that takes the form of an assignment to create collateral, denominating that document an "assignment" rather than a lien, and (2) file a financing statement pursuant to the applicable state U.C.C. statute.

Unfortunately, this secure approach of taking collateral with an assignment, or variations on the approach, may have undesirable tax impacts or other negative features, that might be important. These impacts may arise out of the legal title, if not equitable title, that may be created for the "secured party." For example, revenues from the arrangement may be considered royalties from patents, rather than interest from secured loans, and certain expenses may have to be capitalized and amortized instead of expensed. Also, an assignment as an alternative to a lien may put the "secured party" in the chain of title as a patent holder and licensor, and as such may increase the exposure of this party to tort liability and create obligations for patent maintenance and enforcement.

These issues point out the need to amend the patent statute to correct these deficiencies. A bill containing amendments to correct these problems with Section 261 has been drafted and proposed to Congress. (See the Appendices in this book.) This bill would create a competent statutory title recording regime for patents (analogous to what we have had for a long time for real property and personal property) and interface with existing UCC and bankruptcy priority rules and foreclosure procedures. By providing classic statutory answers to questions of patent title and collateral, the value of patent property will be increased, and it will be easier to finance the development of this type of patentable technology. This is an effort that all in the technology community should support, but to date it does not appear to be a priority in Congress and fast action can not be expected.

Step 10. Confidentiality Program. Determine whether a complete modern confidentiality program is in place in the corporation. This can act to promote the growth of the patent portfolio of the corporation, to increase the value of the portfolio, to keep ownership of all forms of intellectual property within the corporation, and to prevent loss of IP title to the public domain or the hands of competitors.

Step 11. Latent Liability: Full Material Disclosure, and Waste of Corporate Assets. Another question that may be asked is whether the corporation duly exercised full material disclosure regarding its material intellectual property portfolio in prior offerings of securities, both public and private. If there were material failures in disclosure, then these offerings may represent latent securities liability to the corporate issuer.

If a corporation has not adequately developed a patent portfolio for its products and services, it may also be open to liability to shareholders for waste of corporate assets. This may be of particular concern in industries with newly patentable products, such as software and financial investments.

Step 12. Enforcement. The business of competitors can be analyzed to determine if they are infringing the patent rights of the company. If so, opportunities may exist to suppress competition with injunctions and to increase cash flow with judgments and royalties.

Step 13. Revenue Enhancement and Intellectual Property Audits. Many firms under use their intellectual property. An "intellectual property audit" within the company can identify latent intellectual property (patents, copyrights, or trade secrets), that can be protected and licensed to others. In the last few years, some firms have realized material revenues, at substantial profit margins, by programs such as this.

Step 14. Other Intellectual Property. In addition to patents, the analogous steps described above can be taken with the other forms of material intellectual property, including copyrights, trademarks, trade secrets, and confidential information.

Exercising at least some of these steps can be critical to the success of any financial transaction with a technological company, where due diligence and full material disclosure applies.

Three Practical Tips

1. A Competitive Advantage. Some large companies have sophisticated in-house professionals who fully understand their intellectual property and how to deal with it. Most companies cannot and do not have such talent on staff.

Likewise, many operations that are financing technology companies, including some well known venture capital concerns, cannot and do not have the expertise to evaluate a patent portfolio, and its impact on a company.

Furthermore, many patent law practitioners are not aggressive in pursuing these issues, for a variety of reasons which may include lack of client interest, lack of experience, and lack of personal interest.

Consequently, a large competitive advantage can be had in the finance of technology companies by those who obtain access to patent professionals who can add value to a financial transaction.

2. An Epidemic. Basic title problems are everywhere in the patent world. But they can often be cured if acted on early enough. Ownership issues are the first thing to review in a patent portfolio.

3. Budget: Fifteen Items. Almost no deal has a budget or schedule that will allow all possible patent due diligence to be done. The idea is do the basics, and then use judgment and some creativity to react to the basic results and to ascertain what else to do.

For example, validity opinions are rarely done in patent due diligence, because of the customarily large expense.

Perhaps the following is a good minimum menu for patent due diligence:

(1) Properly identify all patents in the key transaction contracts, and use words of conveyance for these patents.

(2) Verify recorded assignments from all the inventors.

(3) Review all Patent Office filings and UCC filings for clouds on title.

(4) Include all foreign patents of interest.

(5) Review all licenses, partnership agreements, employment agreements, and other relevant contracts. Watch out for ownership "leaks" especially for "after developed" inventions.

(6) Do you need a title opinion letter for the patent from your patent counsel?

(7) Ascertain if the target's products fall within the scope of their patents.

(8) How secure are the target's trade secrets? Review its "confidentiality program."

(9) Has the target patented its software or its services?

(10) Review all the target's past attorney opinion letters regarding patents.

(11) Review all past "cease and desist" letters regarding infringement, both sent and received. Has laches arisen? Is infringement willful? Is the target about to be sued?

(12) Investigate the competition's patent portfolio, U.S. and foreign. Get your patent

advisor to characterize the forest, not just the trees. Where is the industry going, and how does the target fit in? Can the target protect its new products and services with patents?

(13) Get representations and warranties from the target, and its patent counsel, regarding all of the above.

(14) Once you have done the above, step back and decide what else should be done in this particular case, within the current schedule and budget. For example, can the target "invent around" its competition's patents? Can the target "invent-on-demand," or "invent-just-in-time," to evolve its currently unpatented products into new patentable variations?

(15) Remember, if you cannot cure a problem any other way, you can always change the prices in the deal, set up escrows and offsets securing representations and warranties, or ultimately (and this is the fire escape of last resort) you can just not close. Litigation, of course, is not a business transaction, and is best avoided.

Discussed elsewhere in this book is a chapter on the personal liability of officers and directors for corporate infringement. These topics closely relate to the due diligence issue for financing technology companies.

Due Diligence as Industrial Espionage

Regardless of the original intent of the parties, any due diligence project can in retrospect appear to have worked like a sophisticated commercial espionage project against a competitor.

Take this scenario. Big Fish Co. expresses an interest in investing in, buying out, joint venturing with, or using as a supplier, the aspiring Mullet Co. Mullet Co. is of considerable interest because it has developed a wonderful new fish hook early warning system, but does not have the capital to market it outside the metropolitan area of Pikes Peak, Colorado. But before Big Fish can make a decision on Mullet, Big Fish must do its due diligence and become an expert on Mullet and its product.

In the due diligence process, Mullet shows Big Fish its books, and the software algorithm that makes the hook alarm so unique and effective. Big Fish loves the algorithm, which is so brilliant that it requires only four lines of source code. But Big Fish walks the deal because its bylaws require that all members of the Board of Directors have a driver's license, and this disqualifies the President of Mullet Co., until he has two more birthdays. And Mullet Co. insists: no Board seat, no deal with Big Fish.

Six months later, Big Fish is on the market across the continent with a hook alarm that works almost as well as Mullet's and costs twice as much. But Mullet Co. is made into fish meal by Big Fish's huge sales force.

Was all this a sophisticated spying job by Big Fish, or an unavoidable development? After all, most prospective deals never happen. In fact, the answer to the question means little to the parties, except as it may effect their legal rights.

The tension in due diligence is that the cooperative target must treat due diligence as an actual espionage activity, but still try to get a friendly deal done. Non-disclosure agreements can be obtained, but they are only contracts that can be breached, and for which any defendant may have defenses. The Uniform Trade Secrets Act may give some recourse, if the aggrieved party survives. Perhaps the best course is for the target to reveal as little as possible, and no technical product information that is not in pending or issued patents.

The lesson of the *Stac Electronics v. Microsoft Corp.* (cited herein) litigation is instructive in this light, and is discussed in more detail in Chapter 12 of this book, "The New Paradigm for Software and Financial Products."

In the above scenario, if Big Fish is in fact using the Mullet algorithm, Mullet will be best protected if the algorithm is claimed in valid patents owned by Mullet. As Stac showed in its litigation, only this strategy can make Mullet into a piranha, instead of chopped bait.

8

Patents for New Business Plans

If you have an idea for a new business, you might engage a lawyer and ask if your concept can be patented. You will probably get the conventional answer: businesses cannot be patented; only devices, processes, and chemical compositions can be patented. You have heard of trick questions; this is, in some cases, a trick answer.

Computer systems and computer applications can be patented. Hardware and software combinations can be patented. And after recent federal cases such as *In re Alappat* (1994, cited herein), it is clear that even pure software is a candidate for patenting. A patent in these areas must meet certain conventional requirements. That is, it must be new and useful and not disclosed or sold in the public domain more than one year before the patent application is filed (or, for foreign patent applications, at any time before the application is filed).

Although new software that meets these requirements is eligible for patents, the patent application must be artfully drafted. For example, in a pure software patent it may be prudent to refer to at least some hardware elements in the claims (even if they are no more than references to using a computer to executing an algorithm), and to storing the results of computer algorithms in memory, or to some other way of producing tangible manifestations of information output.

In highly automated computerized businesses, such as the telecommunications industry, many new concepts can be executed in a practical manner only by application of automated software and hardware systems. Therefore, new computerized business concepts may be monopolized, practically speaking, for the life of a patent that broadly describes the computerized implementation of the business concept. This is very interesting in the telecommunications industry where new technology is emerging and almost all new services are automated (and hence potentially patentable).

For example, recently several patents have been issued involving telephones and faxes in an attempt to lock up new 900 number services and businesses. A few include:

- No. 4,893,333 for "Interactive Facsimile System and Method of Information Retrieval."
- No. 4,974,254 for "Interactive Data Retrieval System for Producing Facsimile Reports."
- No. 4,918,722 for "Control of Electronic Information Delivery."
- No. 4,941,170 for "Facsimile Transmission Systems."

An example of how a software patent can effectively monopolize a business can be seen in Merrill Lynch's Cash Management Account (CMA) patent. The CMA account is a concept invented by Merrill Lynch. In one account with one monthly statement it combines features of an interest-bearing checking account, a security brokerage account, a margin account, a credit card account, and other account features. Normally, one would consider the CMA account a "business method" which is not patentable. However, Merrill Lynch obtained a patent for the *computerized bookkeeping* for this kind of account.

You might think that with current technology it is impossible to offer a CMA-type account on a large scale without bookkeeping by computer. What is one to do, keep the books with a pencil? In the federal case *Paine Webber v. Merrill Lynch* (1983), Paine Webber sued Merrill Lynch to have this patent declared invalid. Merrill Lynch counter-claimed against Paine Webber for patent infringement. The court decided that the patent may be valid and infringed by Paine Webber. Paine Webber settled out of court with Merrill Lynch rather than risk further litigation.

Many activities that are now very profitable segments of the interactive telephone industry may have been candidates for patents when they were first introduced, but no applications were filed then. Unfortunately, once a concept goes on sale it enters the public domain, unless a U.S. patent application has not been filed within one year of sale. (Or for foreign applications, it goes into the public domain immediately, if an application is not filed before it goes on sale.) For example, 900 party lines, 900 fund raising and political contributions, and the concept of distributing information orally from a data base by a 900 number, may all have been eligible for patents when first introduced. Apparently, however, no such patents were pursued, and they lapsed into the public domain for free use by imitators. However, recently, patent applications have been filed, and patents issued, for new narrow 900 number services and businesses.

News is getting out in the telecommunications business on the power of patents for new services and businesses. Some of the most innovative minds and companies in the business are now staking out their patent turf for new opportunities available for the future.

The inventive minds in telecommunications are not going to forfeit their secrets to the competition anymore. Like the Oklahoma Land Rush, they will first stake out their patent claims where possible to their newly invented intellectual property. Then they will do business with a legitimate monopoly on the scope of their patents.

9

Patent Litigation As A Business Tool

"How much justice can you afford?"

Patent litigation is best viewed as a business tool to pursue the commercial goals of the parties. Like any other investment opportunity or operations decision, patent litigation can be analyzed and controlled by application of the quantitative techniques of operations research and systems analysis. A number of non-quantitative strategies of patent litigation are also discussed here, which are quite effective in achieving cost-effective results for a company in a potentially adversarial patent situation.

Patent litigation as a business tool has become increasingly powerful, and dangerous, since the mid-80's. At that time, changes were made in U.S. patent law to encourage the enforcement of U.S. patents. In one example of this increasing power of patents, a recent patent infringement action, *Polaroid Corp. v. Eastman Kodak Co.*, 867 F.2d 1415, 9 USPQ2d 1877 (Fed.Cir. 1989), resulted in a judgment of almost $1 billion dollars and a permanent injunction to abandon an entire line of business.

The literature of patent litigation is arcane and tends to be written by and for specialist patent attorneys. This writing is focused on the technical details of trying and winning patent law suits. Although this is a necessary and useful body of analysis, it is also necessary and useful to expand this analysis to include the broader view of patent litigation as a business tool. In this way, the client and the litigator can collaborate to manage litigation for the broader goals of the client.

It is illustrative of the focus of much of the analysis of patent litigation to review the contents of recent leading seminars for continuing legal education about patent litigation. For example, one such recent seminar by patent attorneys for patent attorneys, was a two-day activity at which almost 2,000 pages of written material was distributed. This material focused on various topics of patent litigation, such as managing legal expenses, preparation of expert witnesses, discovery, use of demonstrative evidence, jury selection, presentation of evidence, measurement of damages, infringement liability insurance, opinions of counsel, injunctions, the validity of patents and trends in legal fees for litigation. However, there were not any discussions scheduled for questions such as settlement strategies and negotiations (even though most patent litigation matters settle without going through trial to judgment), the evaluation and management of litigation from the client's point of view, re-examination options in the U.S. Patent Office in response to an infringement suit (this can be a fast, cheap way to cancel a plaintiff's patent before trial), concepts of inventing around a competitor's patent to avoid infringement litigation, counter-attack strategies available outside of the courtroom to defendants in patent litigation, multi-national patent litigation (most markets in technology products are global but a U.S. patent litigation matter can cover only U.S. markets), benefit-cost analysis (investment analysis) for decision making for ongoing choices in the litigation context, goal analysis for patent litigation, preventive measures to avoid litigation, the use of second legal opinions for the decision to litigate and the conduct of litigation, steps to minimize personal liability of officers and directors for patent infringement issues, and the use of an intellectual property committee of the board of directors to deal with these and other issues.

Operations Research and Patent Litigation

A law suit can be analyzed like any other business investment opportunity or operations decision. Applying quantitative techniques of operations research and systems analysis, together with patent legal expertise, can help control decision making for patent litigation. However, today operations research techniques are generally not applied to control litigation situations. This is probably because litigators, and business people trained in operations research, tend not to be cross-trained in each others disciplines. However, the collaboration of operations research and legal expertise can allow management to avoid abdication of decision making to trial attorneys in a litigation situation, and allow attorneys to be more valuable and responsive to their business clients. The paradigm here is definitely not for the business client to make a one time decision to litigate, and then retire to the role of a spectator and payor of legal fees.

Benefit-cost analysis, and related quantitative analytical techniques of operations research, can analyze the decision options in litigation, as in any other investment decision. Decisions can be clarified by an itemization of the possible outcomes, a structuring of the interrelationships and probabilities of the outcomes, an estimate of their costs and schedules, and an estimate of their financial benefits. This basic approach can force consideration of the full range of decision options, and focus analysis on what net value, or profit, the different choices may have. This in turn provides the basis for objective quantitative comparison of the various decision options available in a litigation situation, and, in the end, better control and better results.

Decision Trees and Risk

As in other areas of business activity, litigation represents an ongoing evolving array of options and probabilities, rather than one digital go or no-go decision. Substantial risk and unpredictability are present. Hence, dynamic decision tree analysis is applicable to litigation decisions. Each possible outcome, should it occur, has some anticipated financial value (positive or negative) to the company and some probability of realization. The occurrence of each event opens up the possibility of other resulting events, and forecloses any possibility of some previously possible events. The probability of a possible event times the value of the possible event if it occurs, represents the expected value of the given event. Discounting the expected value of the outcome at the applicable discount rate for the company from the expected time of realization to the present yields the expected present value of the possible outcome. In this way, a decision tree of the possible outcomes of a litigation situation can be structured, each branch of the tree can be evaluated, and a decision thereby facilitated. In particular, this approach can accommodate some appreciation of the unpredictable nature of litigation, and the fact that some outcomes can resolve the situation much faster than others.

It is useful here to show the general structure of a decision tree analysis in a patent litigation opportunity. In the most general sense a patent infringement suit by a plaintiff against a defendant has several classic possible outcomes. The plaintiff may obtain a injunction against further infringement. The plaintiff may obtain a cash payment of a certain amount. The payment may be tripled for willful infringement. The plaintiff may also obtain an award of its fees and costs. The plaintiff may also experience considerable expense and obtain no injunctions or payments. The defendant may also counterclaim against the plaintiff, claiming that the plaintiff's patent is invalid, and seeking a ruling of invalidity against the plaintiff's patent. The defendant may also claim that the patent was obtained through inequitable conduct, and obtain a ruling that the patent is unenforceable. The defendant may file a request for re-examination of the patent at the U.S. Patent and Trademark Office, which may result in a cancellation of patent without the time and expense of a trial.

A quantitative decision tree analysis of a patent litigation opportunity would include quantification of these general possible outcomes applied in the specific context of the facts of the case. Monetary values of the different outcomes would be determined by the business context of sales, market prices, and profit margins, and by the rules of law used to quantify certain litigation awards. The probability of each outcome would include an assessment of the strength of the legal case, the strength of the participants in their business, and the assessment of the actions of the other side. The economic impact of each outcome would vary for each party to the case, so that a separate evaluation would be needed for each party, even though the range of outcomes and probabilities of the case would generally be the same or similar for each party. These quantified estimates of economic impacts can best be made by the business clients in consultation with their legal advisors.

To illustrate the decision analysis technique applied to patent litigation, a hypothetical case is presented in the four accompanying figures.

Figure 1 shows a decision tree for a hypothetical case that uses a basic structure applicable to all patent infringement litigation. This figure is presented from the defendant's point of view. (A similar figure, but from the plaintiff's point of view and with different values, is developed in much the same way in Figure 3.) In Figure 1, the basic decision to license (that is, settle) or litigate at the beginning of the process is emphasized. If the defendant settles with a license from plaintiff, the estimated present value of the cost of royalties in this hypothetical is $3,000,000, and the estimated present value of the future profits from sales under license (before royalties) is $4,000,000.

In the event of litigation, the possible determination of the questions of patent validity, infringement, and willful infringement are shown. The estimated probability of each outcome is shown, together with the estimated cost of each outcome to the defendant (shown in units of millions of dollars). In this example, the cost of legal fees and costs to the defendant of going to trial is estimated at $1,000,000. Hence, if the defendant goes to trial and wins, either by invalidating the patent, or by showing no infringement, the win costs the defendant $1,000,000. If the defendant loses (the patent is upheld and infringement found), then the defendant pays an estimated $8,000,000 if no willfulness is found (legal fees and costs of $1,000,000, the cost of the permanent injunction estimated at $1,000,000, and the cost of actual damages, estimated to be $6,000,000). If willfulness is found, the defendant pays $20,000,000 (with triple damages for infringement). No discount for the value of time is used in Figure 1, except for the present value of sales if there is settlement.

In Figure 1, the net expected present value of licensing is $1,000,000, versus the expected cost of litigation of $6,900,000. In this example, settlement is indicated for the defendant.

Figure 2 shows the same analysis as Figure 1 for the hypothetical case, except that the value of each outcome is discounted for the value of time at an annual interest rate of 15%. It is assumed for this example that the initial license would settle immediately at the indicated present value. It is assumed that the other outcomes would occur after four years of litigation (an annual discount rate of 15% compounded over four years results in a discount factor of 52%).

Figure 1
Basic for defendant – no discount for time

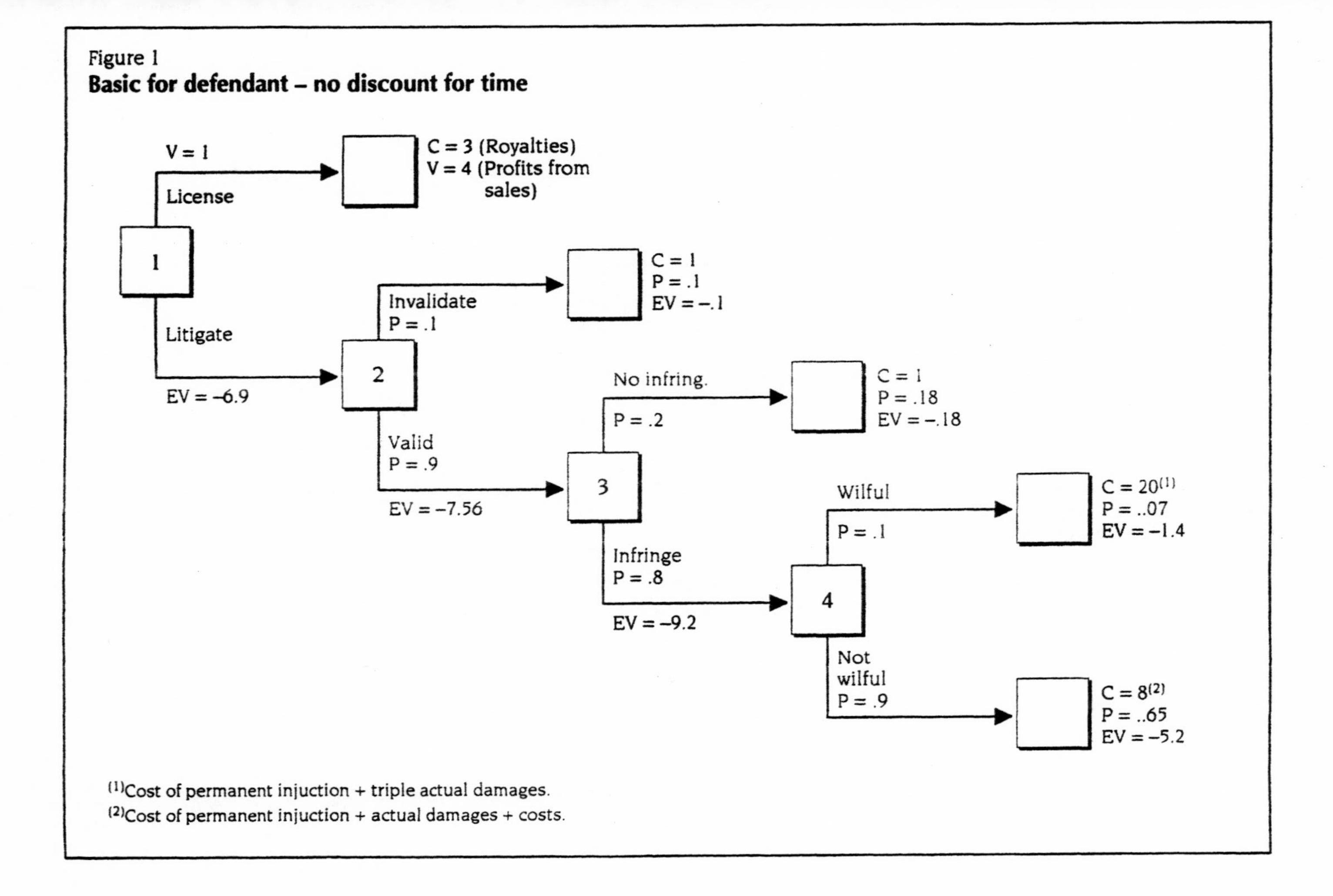

[1] Cost of permanent injuction + triple actual damages.
[2] Cost of permanent injuction + actual damages + costs.

Figure 2

Basic for defendant – discounted for time (4 years to trial at 15% interest rate)

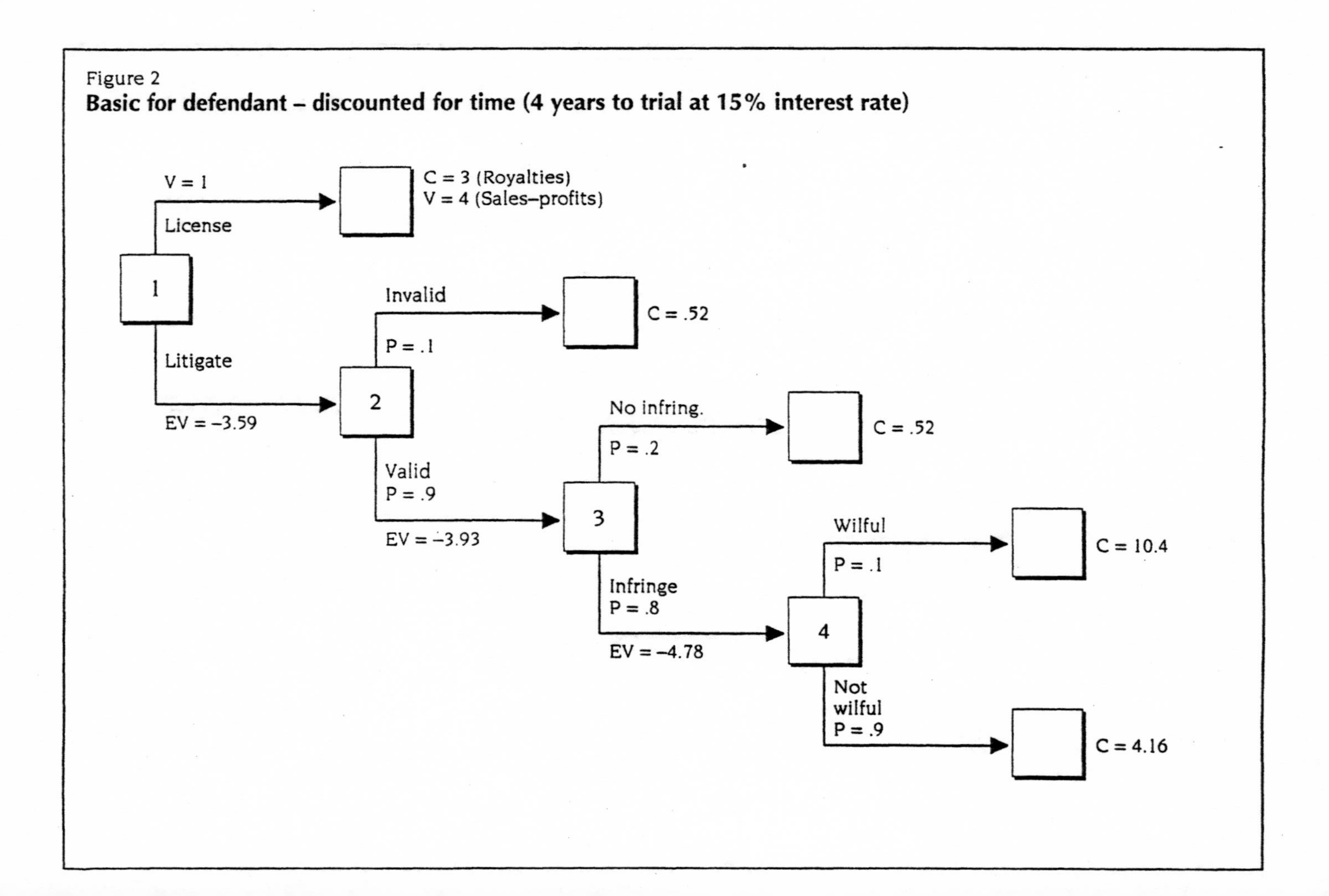

Figure 3
Basic for plaintiff – no discount for time

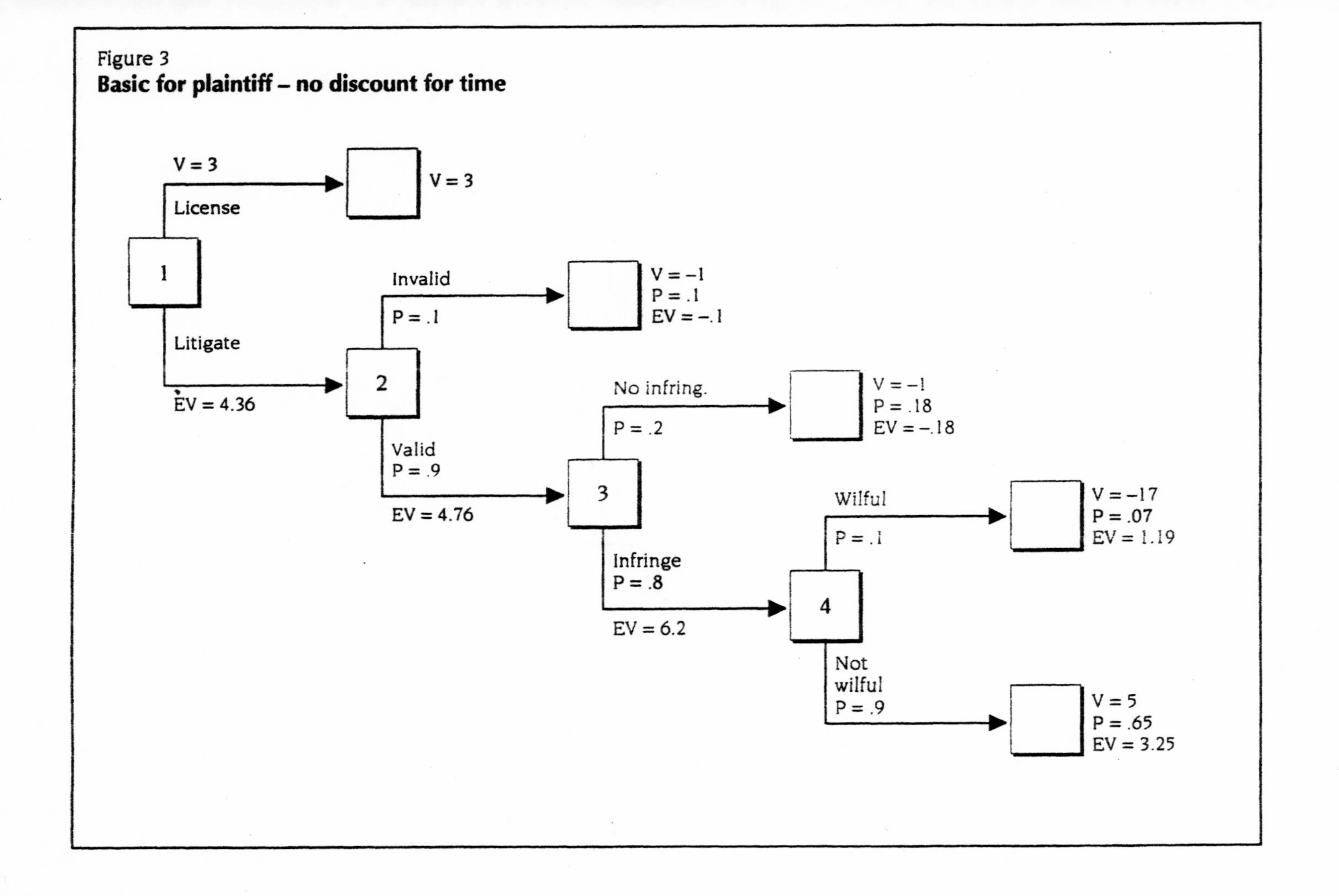

Figure 4

Basic for plaintiff – discounted for time (4 years to trial at 15% rate)

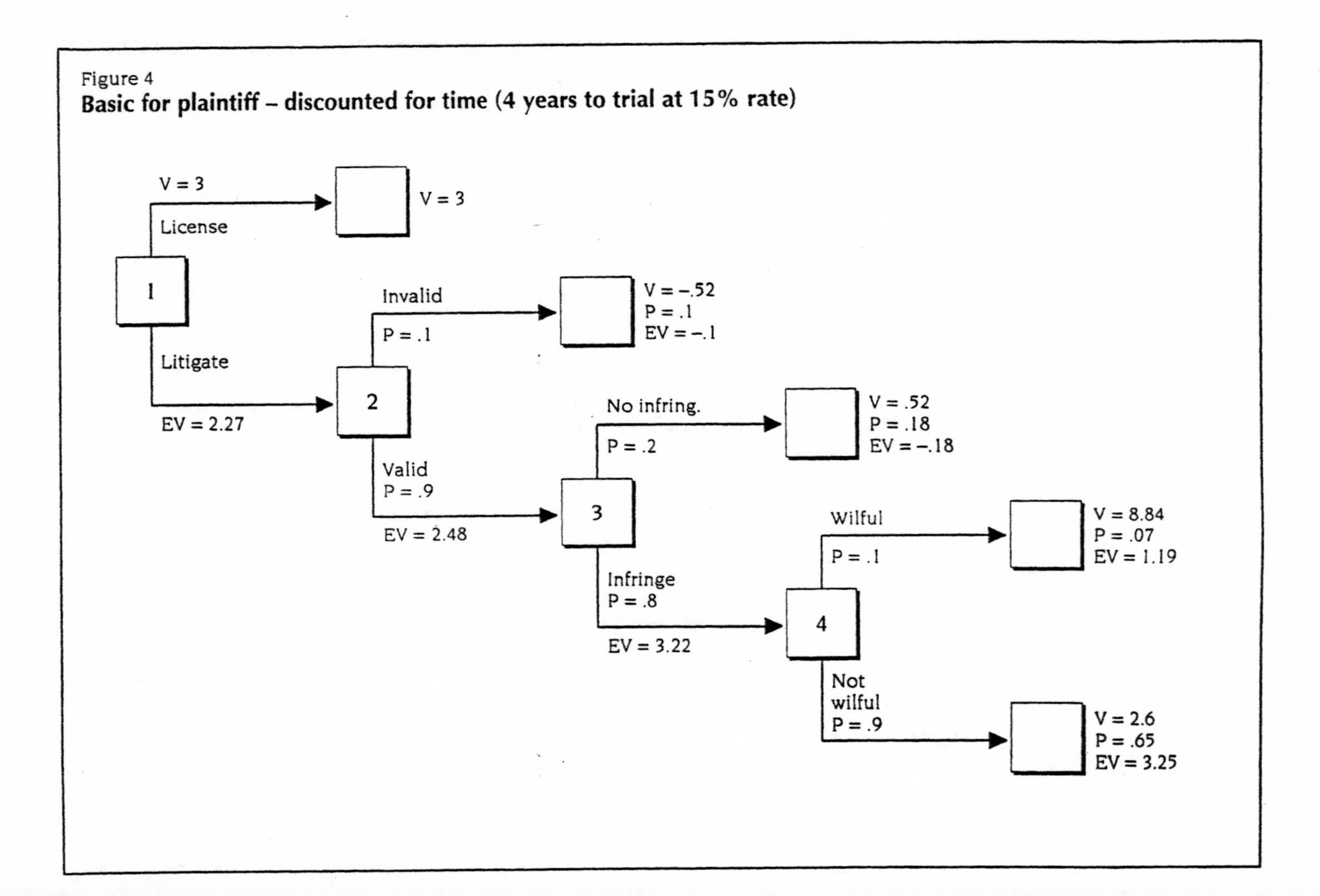

In Figure 2, with adjustments for the time value of money, the expected present value to the defendant of settling is $1,000,-000, versus the present expected cost of litigation of $3,590,000. Settlement is still indicated to the defendant. However, because the larger costs of litigation are in the future, whereas the smaller net present value of settling initially is in the present, it is a closer decision for the defendant to settle when the time value of money, and the delaying effects of litigation, are considered.

Figure 3 shows an analysis of the same hypothetical case in Figure 1, but evaluated from the plaintiff's point of view. The present value of settlement to plaintiff is estimated at $3,000,000 from the royalties. If the patent is litigated and invalidated, the cost in legal fees and costs to plaintiff is estimated at $1,000,000 (assuming no lost sales of plaintiff as defendant continues in business). If trial is had and the patent is invalidated or no infringement found, then plaintiff loses $1,000,000 in fees and costs. If the plaintiff wins, but without a finding of legal willfulness or an award of attorney fees, then plaintiff gets $6,000,000 in actual damages, less $1,000,000 in costs (with no value in sales attached to the injunction in this case). If willfulness is found without an award of legal fees, then plaintiff nets $17,000,000 (triple damages, less legal fees). Here the expected value to plaintiff of settling is $3,000,000, versus $4,360,000 for litigation. This marginally favors litigation for the plaintiff, but considering the uncertainty of litigation, plaintiff may prefer settlement in this case.

Figure 4 shows the same analysis as Figure 3, except that the value of each outcome is discounted for the value of time at an annual interest rate of 15%. Hence, Figure 4 is the same analysis as Figure 2, except from the plaintiff's point of view.

With the time value of money considered, the present expected value of litigating and winning with triple damages, for plaintiff, falls to $8,840,00. Without willfulness, the present value of winning is only $2,600,000. Hence, the expected present value of all litigation options is $2,270,000 for plaintiff, versus $3,000,-000 for settlement. In this hypothetical, therefore, plaintiff receives a strong settlement signal, especially since the bird in the hand (settlement) is more valuable than the two in the bush (litigation). Coupled with defendant's settlement signal on the same probability estimates in Figure 2, the objective foundation for settlement exists in this case, if the two parties evaluate the parameters of the case in about the same way.

The most interesting result of comparing the indications of Figures 2 and 4 in this example is that it is most advantageous for both the plaintiff and the defendant to settle. In such a situation, settlement is most likely to occur. Indeed, the reality is that most patent litigation situations result in some sort of settlement before trial is completed, appeals are exhausted, and a final judgment is collected.

Note that the more realistic analysis includes the present value discounting of future outcomes. In reality, time is money. Hence, Figures 2 and 4 show the better analysis. The analysis without present value considerations, shown in Figures 1 and 3, is really an intermediate step in preparing the final analysis of Figures 2 and 4, and is shown here primarily to illustrate clearly other necessary principles without present value complications. For purposes of actual decision making, it is simply not realistic to see a dollar today as having the same value as a dollar four years from now.

Further details could be added to the analysis and explicitly incorporated as branches in the decision tree with quantified probabilities and present values, as may be germane in any particular case. For example, analysis of the possibility of a permanent injunction, but with no damages, could be included.

The possibility of an early and successful re-examination of the plaintiff's patent, that results in cancelling the patent (and the plaintiff's case) before the time and expense of trial, could be included. The possibility of settlement after a partial successful re-exam that amends the patent, or after a failed re-exam, could be added. Different settlement strategies during various points in the trial and appeal process could be specifically incorporated into the analysis, as they appear relevant. Contingency fee options may be included in the analysis, to analyze the dramatic effect that contingency fee arrangements can sometimes have on the best way to manage a litigation situation, and especially on the decision of an underfunded plaintiff to litigate rather than settle.

Sensitivity Analysis

In the hypothetical, the probability that plaintiff will win in some manner at trial is estimated at 72%. This estimate, of course, should be based on the facts in each case. It is not surprising that where the defendant has only a 28% chance of winning in trial, the best course would be for the defendant to take a license that allows it to profitably continue in business, if it has the chance.

If, however, for example, the defendant had a larger estimated chance of success at trial and no option of a settlement with a profitable license, then the analysis would more likely indicate litigation for the defendant. Sensitivity analysis can identify the estimate levels at which the defendant's preferred course would switch from settlement to litigation, so that those levels can be reviewed for their likelihood. Sensitivity analysis can also indicate which estimates are the most important to the outcome, so that the most effort can be put into influencing the most important parameters.

Quantitative Analysis of Delay by Litigation

It is an interesting quirk in the quantitative benefit cost analysis of litigation decisions that the value of time may be accounted for in a manner different from most other investment opportunities. In most situations, the present value of a future payment or receipt is discounted by the time cost of money to obtain the present value. That is, a dollar in the future is worth less than a dollar today, by an interest rate equal to the cost of borrowing money of the party in question or the opportunity cost equal to the yield that may be earned in an alternative investment. However, in litigation, one goal of litigation may be delay itself. For example, one party may delay the initiation of marketing of a competing product by a competitor by suing that competitor for patent infringement, thereby delaying the competing product introduction until the resolution of the infringement suit. Hence, the additional delay of litigation would have a component of positive value for one party. In the benefit cost model adjusted for present value, this would not impact the discount rate for the incidence of future costs and receipts. However, it would add an additional component that would increase the present value of any outcome, provided the outcome would delay the competing product introduction until the litigation outcome was reached. The value of this delay effect would be equivalent to the stream of increased profits for the duration of delay, which increase would be caused by the ongoing benefit of avoiding competition from the delayed competing product for the duration of the ongoing delay, and the profit stream would be discounted to present value at the applicable discount rate.

Likewise, where the effect of delay was positive for one party, it is usually negative for the other party. For the party that suffers from delay, the same basic analysis would be pursued, but in place of a stream of increased profits from delay, a stream of costs discounted to present value would be used. Delay would have a negative impact, for example, where the plaintiff is a small patent holder and the defendant is a large infringer capable of eliminating the patent holder in the market place, in the absence of an injunction to cease infringement.

Settlement Scenarios

Statistics indicate that the majority of patent litigation situations settle and do not go to trial, appeal, and collection of a final judgment. Consequently, a complete quantified analysis of the opportunities may include settlement scenarios, which may take place anytime before or during litigation, upon the agreement of the parties. Settlement in lieu of litigation, prior to long and expensive discovery, may be an attractive path to pursue, particularly in light of the early elimination of uncertainty that it may represent. These techniques can facilitate the evaluation of settlement opportunities to ascertain if they can be expected to better serve a party than the litigation alternatives.

Re-Examination Scenarios

Re-examinations are relatively fast and cheap compared to federal litigation and can be pursued anonymously by the party that initiates the re-exam. A re-exam can even be initiated anonymously by a licensee of the challenged patent. Also, a patent being re-examined benefits from no statutory presumption of validity. (A patent in litigation in federal court benefits from a statutory presumption of validity, and the burden of proof is on the challenger, that is, the infringer, to show that the patent is not valid.) Since an early and successful re-examination may be considerably faster and cheaper than a full scale trial process, an early re-examination is a defense strategy that should not be overlooked by the defendant. (Note that a re-exam can be initiated during a trial, and usually will stay the trial pending the result of the re-exam. It appears that most patents that are involved in litigation end up being re-examined, but this tends to happen only late in the trial process, or after trial pending appeal.)

Settlement for Foreigners

The basic settlement scenario has a particularly interesting variation in the case of a foreign patent holder pursuing a domestic U.S. infringer. A foreign patent holder may have little U.S. presence but be interested in penetrating the U.S. market. In this situation, one possible strategy for the foreign patent holder/-competitor would be to develop a U.S. joint venture partner for the patented products or services, where the partner has existing U.S. market penetration. A domestic U.S. infringer would, of course, already have a U.S. market position. Therefore, a natural settlement opportunity would be for the foreign patent holder to reach a joint venture arrangement with the U.S. infringer, whereby the U.S. infringer would be the U.S. distributor or joint venture partner paying royalties or other payments to the patent holder. With such an arrangement, both parties would avoid the expense and uncertainty of litigation and the possibility of total defeat in court (that is, invalidation of patent rights for the plaintiff, or enforcement injunctions and payments of damages for the infringer).

Second Opinions for Litigation

In major litigation, some parties have begun to regularly use a continuing second legal opinion regarding the course of the litigation. In this arrangement, second legal counsel is used in collaboration with lead litigation counsel to review and consider the strategy recommended by lead litigation counsel for the client. Both counsels report directly to the client. This practice may have been first used extensively by Japanese patent holders in the U.S. as a conservative approach to a new legal environment for these parties. In the context where patent litigation is potentially a life or death manner for the parties, it may be judicious in most instances for the parties involved to obtain a second opinion throughout the patent litigation management process, in much the same way that a medical patient may obtain a second opinion regarding important life or death recommendations regarding medical surgery. The evaluation of litigation situations can have very subjective components, and a wide range of opinion can develop even among experts. Also, litigation, because it is adversarial, can be quite polarizing for the participants and lead to a lack of sensitivity to options that may be objectively interesting. In particular, a second opinion regarding settlement opportunities, or retention of a second legal group to pursue settlement in collaboration with litigation counsel, may be fruitful in major litigation.

Contingency Litigation

Contingency litigation is also an option available in some instances for parties in patent litigation. Contingency arrangements provide an alternative to paying attorneys a guaranteed hourly rate for litigation regardless of result. There are many types of contingency arrangements, but their common theme is to have the attorney share some of the risk of litigation to the extent of making at least part of the attorney's compensation dependent upon the outcome of the litigation. Contingency arrangements might involve (1) a reduced hourly rate to the attorney plus a certain percentage of any recovery, (2) no hourly rate but a reimbursement of expenses and costs to the attorney, plus a larger percentage participation in any final judgment, or (3) some form of the preceding arrangements, with specific cash bonuses for specific outcomes (such as settlement before trial, or obtaining a permanent injunction). Contingency arrangements are possible with both plaintiff's and defendant's counsel. These arrangements reduce the guaranteed out-of-pocket costs to the parties and provide performance based compensation in some measure to the litigators. It also adjusts the cost of litigation to a degree to be proportionate to the value received by the client. This situation has been encouraged by the trend toward enforceability of patents in the United States, and the possibility of obtaining triple damages for willful infringement. (This might be referred to as Glazier's First Law of Litigation: wherever a statute provides the possibility of triple damage, then a contingency bar arises to conduct litigation in an entrepreneurial spirit.)

Inventing Around

Another possible outcome of patent litigation may be that the defendant invents around the plaintiff's patent position. That is, the defendant invents and patents a new product or service for the same market, which new product or service does not infringe on the plaintiff's patent, and which new patent the plaintiff itself must not infringe. It is possible in some circumstances for a defendant/competitor to avoid a valid patent in this manner, at least for the future. This approach is especially effective where the defendant can also invalidate the plaintiff's patent.

A patented product or service serves a particular market niche, however, that same market niche may be accessible to an alternative and non-infringing product or service. That alternative product may even be economically fungible or superior to the patented product from the customer's point of view, even though that alternative product may does not infringe upon the patentee's patent. To develop such a non-infringing competing product, that is to invent around the patent holder's patent, is one possible response of a defendant to patent infringement litigation. The invent-around response by a defendant may also lead to an eventual infringement counter-attack by the original defendant. This counter-attack may result where the invent-around response by the original defendant leads to a patent portfolio of the defendant's own which may be used to attack later second generation developments by the original patent holder.

Goal Analysis

The selection of different possible outcomes during the course of development of a litigation situation may be influenced by the preferred goals of the parties. Litigation is an ongoing evolving process for which decisions are constantly being made. These decisions tend to favor outcomes in certain preferred directions. Consequently, if a company is concerned with maintaining a one-producer market, then the only goal that has much value for the plaintiff may be a permanent injunction against future infringement. However, if the patent holder merely wants to optimize its receipts or income, then the patent holder has a wider range of outcomes to elect to pursue. This range of possible outcomes may include a judgment of infringement and damages, settlement involving license and royalties, or otherwise. Also, if a patent is thought to be weak and susceptible to attack by re-examination, then the patent holder would see a larger probability of invalidation or cancellation of the patent. This would lead the patent holder to favor early settlement at reduced costs, thereby avoiding damage to the patent and establishing a precedent of enforcement and a royalty level.

Coordination of International Litigation

U.S. patents and patent litigation provide protection and injunctions that are effective only in the U.S. In a global marketplace with the possibility of global patent strategies and global infringement, patent litigation must take a coordinated international view. This is often the case, since in these situations, more or less the same case, with different details, is often litigated repeatedly against the same group of infringers in the national courts of the various major national markets.

In particular, it is often advisable to have an overall managing counsel for related international patent litigation, especially when the litigation involves related facts and defendants in various national forums. This is especially true when the same or related legal questions and matters of fact are to be re-litigated in several national forms, that is, in cases where the record in an earlier forum in one country may be relevant and have some estoppel impact in subsequent forums in other nations. Also, global coordination may be necessary to prevent a party from taking contradictory positions in successive national forums. When litigating against the same or related parties in various national forums, any alternative conflicting positions by one party may be especially likely to be discovered and presented in the subsequent forums.

Prophylactic Measures and Personal Liability of Officers and Directors

Perhaps the best prophylactic measure to avoid infringement is to obtain a review of the infringement question and a written opinion of non-infringement by independent patent counsel, regarding any new product of service, prior to the market introduction of that product of service. In the U.S. such a prior opinion of non-infringement, when provided in writing by an independent patent counsel, may eliminate the tripling of damages in those instances when liability for infringement is eventually found. Such an opinion might also provide some prophylactic effect against any personal liability, both civil and criminal, of officers and directors for corporate infringement, where such infringement is eventually found.

In industries that are very sensitive to technology, it may be judicious for the board of directors to form an intellectual property subcommittee. This is similar to the way that some boards have formed audit subcommittees, compensation subcommittees, or real estate subcommittees to monitor and minimize exposure to certain possible liabilities and to exploit certain classes of opportunities. An intellectual property subcommittee of the board of directors could oversee the effort of the corporation to maximize the development and exploitation of its intellectual property estate, and to oversee the effort of the corporation to avoid large judgments and permanent injunctions for patent infringement. Also, officers and directors of corporations may suffer personal liability for corporate infringement, so the matter is of considerable potential interest to the individual members of the board of directors. Liability to corporations and their agents, both civil and criminal, may arise from patent infringement itself, collateral failure of full material disclosure in a corporate securities context, violations of certain related criminal statutes, and from shareholder derivative claims regarding waste of corporate assets.

Total Quality Management ("TQM")

A lot has been written recently about TQM ("Total Quality Management") and the subject remains controversial. Perhaps it is still an open question in some circles exactly what TQM may be and what it is worth. However, the best discussion at this time about TQM applied to litigation may deal with themes of quality control, economic management of results, and measurement of goals. Applying the operations research techniques discussed above to patent litigation provides an objective quantified structure for the analysis of patent litigation, and thereby facilitates the best goals of TQM.

Indeed, the objective quantitative structure that operations research can give to the litigation environment can give legal managers, and non-legal business managers, a means to control and direct important activities in the difficult and arcane patent litigation area. Lay businessmen, using operations research techniques, need not abdicate evaluation and decision making in patent litigation to expert patent litigators. This allows business goals and judgments, which may generally be alien to experts in patent litigation, to be brought to bear on hostile patent disputes. Hence, the power of patent litigators can be harnessed to best pursue the business goals of their clients in a collaborative effort with the clients. Important litigation decisions do not have to be made in a vacuum by litigators operating in an arcane specialist's world.

The Evaluation of Litigation, Settlement, Arbitration and Alternative Dispute Resolution ("ADR")

The techniques illustrated here give a common framework for both parties in a patent dispute to evaluate the value of their respective positions, in an objective manner. This permits the communication and common understanding necessary and desirable for a consensual settlement of a dispute, arbitration, or Alternative Dispute Resolution ("ADR"). Most of all, being quantitative, these techniques provide a method for the parties of a dispute to calculate one number as the value of their respective positions in a dispute.

Hence, these operations research techniques allow the parties to a dispute to model and quantify the expected objective results of litigation. This gives the parties the opportunity to settle for the expected value of the litigation, as an evaluation of their current positions. Ideally, the parties may reach a consensual settlement without the time and expense of the litigation itself. In other words, in the best case if the parties can quantify the expected results of the trial in advance, they may settle on that, and eliminate the trial.

Of course, operations research can only facilitate settlement in conditions where settlement is otherwise possible. Operations research can not be expected to lead to settlement processes, where the parties would otherwise not permit them. As discussed herein, these quantitative techniques can be applied to evaluate the settlement alternatives, and indeed may point away from settlement to litigation, where the facts so warrant. Operations research can only better analyze the facts, it can not change the facts to make them better.

Portfolio Theory and Patent Litigation

Portfolio theory categorizes investments in different types, each type with its own parameters, and then suggests that investment funds be spread over different types of investments to achieve the best long-term performance.

The application of portfolio theory to the view of patent litigation as an investment opportunity can lead to the identification of categories of patent litigation, categories of parties to patent litigation and categories of patents. Portfolio theory can also lead to the better understanding of classes of patent litigation situations, and to better strategic decisions about patent litigation.

1. Categories of Patents

Patent claims can be broad and fundamental, or narrow and specific to one product. Broad patents can be very valuable because they can cover a range of valuable products. These "market patents" also can be more difficult and expensive to obtain and to enforce, because there can be a broader range of prior art that can block or invalidate them.

Narrow patents can be very specific to one particular product, and not inhibit all competing products in a market niche competing for the same customers. Consequently, the value of these "product patents" can be limited by the value of their one product (although that may be substantial). These patents can be cheaper and easier to obtain and enforce, partly because there may be less prior art that can block or invalidate them, and because the question of what constitutes infringement can be clearer.

Better patents, and larger patent portfolios, tend to have a mix of broad and narrow claims. Pursuing a strategy of patent protection for a market, a product line, or a product should include consideration of pursuing a mix of broad market patents, and narrow product patents.

2. Categories of Patent Litigants

Parties to patent litigation can be larger or smaller entities. Large entity patent litigants tend to be well financed, capable of sustaining long and expensive litigation, able to sustain total defeat in court without fatal financial impact, and less likely to see in a compromise an improvement in their previous condition.

Small entity patent litigants tend to have less financial strength, less ability to support long and expensive litigation, are more likely to be put out of business by a litigation defeat, and are most likely to see a compromise settlement offer as an improvement in their previous circumstances, or even as a goal in their original business plan.

Consequently, in a David-and-Goliath mismatch between a large entity and a small entity, it is more likely that David will want to settle, but does not have the chance, and that Goliath will want to fight to the death, and can choose to.

3. Categories of Litigation

Certain types of patent litigation opportunities may have higher costs and risks. For example, litigation with a large adversary concerning broad market patents may have the potential of requiring large amounts of time and financing for a defendant.

Conversely, some patent litigation opportunities may be less risky. For example, litigation with a small under-financed adversary, over a narrow product patent, may be less threatening to the plaintiff.

To the extent one can control the situation, it is of course preferable to litigate the high risk opportunities only when the pay-off is large, and involves a very valuable patent. Less threatening litigations, however, might be reasonably undertaken for potentially less valuable patents.

Application of decision analysis, as indicated herein, should clarify whether high risk situations are indeed high reward situations, and quantify the decision.

This view also suggests that where a patent holder is contemplating enforcing a patent against a variety of defendants, the patent holder might best act first against the weakest infringer. This might then establish a precedent of validity and enforcement against a weak adversary, making later moves against stronger parties somewhat less risky.

Furthermore, a strategy is suggested, especially in the case of a weak patent, that settlement be offered at royalty rates a bit lower than the cost of a successful defense by the defendant. Hence, the defendant comes out ahead, and the plaintiff realizes cash flow from an untested patent that might not survive litigation.

A Word About Quantification

It is not possible to reliably predict the outcome of litigation. It is also probably never possible to know if one's estimates of the probabilities of events were accurate. However, it is certainly possible to accurately develop the structure of germane possible outcomes in a decision tree for a litigation situation. This in itself can be very useful for management to understand and participate in the control of important litigation. Furthermore, the quantification of estimated probabilities, costs, and revenues for such a decision tree, even though imperfect, can improve everyone's understanding of the situation. This is particularly true when the limits of quantification are realized, which can be aided by sensitivity analysis.

In any case, this approach is an excellent path to integrate a client's business goals with a trial specialist's expertise, to jointly manage a litigation situation to best pursue the business goals of the client.

Litigation in General as a Business Tool

The discussion here emphasizes particular examples in patent litigation. However, the operations research techniques and the point of view used here are generally applicable to all litigation.

All litigation should be viewed as a business tool, not just patent litigation. And the quantitative techniques of benefit cost analysis, present value, probabilities, and decision trees are equally applicable to evaluate and capture the opportunities of any litigation for the business client. In this way, any potential litigation situation can be better made into a business tool to pursue the commercial goals of the client.

A Note on the Evils of Litigation

The non-judgmental manner in which this book deals with litigation should not be taken by the reader as an acceptance or promotion of litigation as it exists in this country today. Indeed, the author feels that there is too much litigation, and too much legal expense in general, in our economy. Numerous reforms, or revolutions, in the system are required.

Our business environment suffers from a sort of economic arteriosclerosis from too much litigation, too much legal risk, and bad rules of law. In general, a lawyer should try to keep a business client out of the adversarial negative sum game of litigation, and in the positive sum mutually consensual game of contractual deal making and smooth business operations.

However, reforming the legal system is outside the scope of this book. What we are trying to do here is to deal effectively with the system as it is, in a client-oriented way. In the short-term, the litigation environment has to be dealt with like bad weather: you cannot change it, so be happy if you can just predict it and adapt to it effectively.

In the long term, of course, we must change the litigation situation and the laws that shape it. Especially, we must consider such tort reforms as (1) limiting class actions, (2) limiting contingency fees, (3) requiring the loser to pay the winner's costs in litigation, (4) returning tort law to the classic concept of actual damages from the proximate cause for negligence, and eliminating the new concept of absolute liability of the proximate deep pocket for the cost of any "victim's" self-destructive stupidity, (5) limiting punitive damages, hedonistic damages, and other baloney damages of whatever name, and (6) generally defending the economic rights of private property in the Bill of Rights to control Robin Hood judges and juries. In particular, we should follow and strengthen the takings clause, and the excessive fines and penalties clause of the Bill of Rights.

10

Personal Liability of Officers and Directors for Corporate Infringement (Civil and Criminal)

"My best ideas are somebody else's."
-Benjamin Franklin

Recent amendments to the federal copyright statute, and a new case regarding patent infringement have increased the exposure to personal liability (civil and criminal) of officers and directors for corporate infringement. The cases also show what steps to take to avoid such personal liability.

Individual officers and directors of corporations can be held personally liable for civil and criminal penalties (including jail sentences) for infringement by their corporations. Individual officers and directors can suffer civil penalties, including triple damages for the corporate infringement of patents. They can also suffer criminal penalties, including jail sentences, for infringement of copyrights (which is now a federal crime, and in many states, a state crime). They can also suffer civil and criminal penalties for violation of the federal state securities statutes regarding any failure of full material disclosure about intellectual property assets during corporate financing transactions involving the public or private sale of securities. They can also suffer individual liability for waste of corporate assets for a failure to properly establish and maintain corporate ownership of their intellectual property assets. (This is especially tricky because the current federal statutory regime provides only a slippery, vaporous form of ownership for intellectual property, which can be easily lost.)

Fortunately, there are a series of prophylactic measures that can be taken with independent intellectual property counsel that may insulate officers and directors from personal liability for corporate mistakes and misdeeds regarding intellectual property. In particular, certain types of written opinion letters of independent patent counsel may serve to free officers and directors from exposure to personal liability for triple damages in civil cases and for jail sentences in the case of criminal infringement. Furthermore, similar procedures are available regarding full material disclosure and waste of corporate asset risks.

Triple Damages for Patent Infringement

Officers and directors of corporations can be personally liable for triple damages for willful infringement of patents by their corporation (35 U.S.C. 284 and *3M v. Johnson and Johnson*, 976 F.2d 1558, 24 USPQ2d 1321 (Fed. Cir. 1992)) and for damages for copyright infringement (*Data Products v. Reppart*, U.S. Dist. Kansas, Nov. 29, 1990, Lexis 16330). This personal liability for damages can accrue even when the individual officers and directors had no knowledge of the infringement and did not personally benefit from the infringement, (*Data Products v. Reppart*, above) and in fact were <u>advised by in-house corporate counsel that no infringement was occurring</u> (*3M v. Johnson and Johnson*, above).

Triple damages accrues from willful patent infringement where the duty of care not to infringe an issued U.S. patent was not fulfilled. The duty of care should be satisfied for any new product to service of a corporation prior to the corporation putting the product or service on sale. (Note that services and processes may be covered by patents in the same manner as tangible products. Software and genetically engineered lifeforms may also be covered by patents.) The duty of care is satisfied by the corporation obtaining a written opinion of non-infringement from independent patent counsel. This sort of prior opinion letter can serve the prophylactic function of insulating the individual officers and directors from triple damages for patent infringement. The details of such an opinion letter depend on the individual circumstances in each case. To insure officers and directors against personal liability for triple damages for patent infringement, in any particular case, the details of the situation should be discussed in confidence with independent patent counsel.

Criminal Liability for Infringement Crimes

The federal copyright statute was amended in November 1992 to make copyright infringement a federal crime, including the infringement of computer software (see, 18 U.S.C. 2319). Furthermore, infringement of any intellectual property has been a crime for some time under various state laws (see, *Schalk v. Texas*, 823 S.W. 2d 633, 21 USPQ2d 1838 (Tex. Ct. Crim. App., Oct. 2, 1991), and Section 31.05(a)(4) Texas Penal Code).

Criminal culpability usually requires the intent of the criminal party. However, for copyright infringement the issue of intent may involve the difficult question of what constitutes infringement, especially in the case of computer software and derivative works in software. In such cases, a written opinion of non-infringement from independent intellectual property counsel prior to the fact may be adequate to establish the lack of intent of individual officers and directors necessary for a crime. (Note that 18 U.S.C. 2319, as amended, is a new statutory provision, with no interpretive cases at this time. However, one can anticipate this trend in case law development, especially in light of the status of civil patent infringement case law.) Again, the details of the content of such an opinion letter of non-infringement from independent counsel, and the effectiveness of the same, would vary with the particular circumstances of each case. Interested officers and directors should discuss the applicable circumstances in confidence with independent patent and copyright counsel.

As is the case with civil infringement, criminal infringement may arise from making, using or selling the potential invention or copyrighted works. This infringement may arise when the manufacturer sells the invention or work in competition with the owner of the patent or copyright. However, this infringement may also arise when a corporation buys or otherwise uses a patented or copyrighted invention or work. This may include software, which can now be protected by both patent and copyright, and which is easily copied.

Full Material Disclosure and Due Diligence

When a company sells securities, the directors and some officers of the issuing company have an absolute obligation of full material disclosure regarding the company and the investment (15 U.S.C. 77k(a)(2)). In the case of a technology company, the intellectual property portfolio of the company is probably a material disclosure item. However, disclosure in American prospectuses is often inadequate regarding intellectual property rights, and deficient when compared with analogous standards of disclosure for real property and personal property. In particular, regarding the products and services made, used or sold by the company, an opinion of non-infringement regarding the patents of others by independent patent counsel is often warranted. Regarding the patent, copyrights and material trademarks of the company, an opinion of validity by independent patent counsel is often appropriate. Regarding pending patent applications, an opinion of patentability is often appropriate.

Regarding the products and services made or sold by the company, an analysis by independent patent counsel regarding the extent to which the markets of those goods and services are protected from competition by the company's intellectual property portfolio is essential. The primary business purpose of patents, copyrights and trademarks is to suppress competition from entering the markets of the company. However, if the patents of the company are valid but simply do not claim the products, then the products are not protected by the patents and the patents may be useless as a business tool. Furthermore, if the patents are extremely narrow, even if they cover the exact products and services of the company, they may still permit a competitor to easily enter the company's market with products and services that are essentially fungible to the user, but which are adequately differentiated from the patents themselves so that they are not infringed. Analysis and opinions of this sort of "coverage" of products by patents is a technical legal function, but it is essential to the evaluation of a technology company. Too often in prospectuses, it is simply implied, without due diligence, investment caveats or analysis, that an issued patent has some value to protect the company's product line from competition.

Similarly, the due diligence defense (15 U.S.C. 77k(b)(3)) would require professionals involved with the financing to inquire into these same disclosure issues and determine their status.

Elsewhere in this book is a chapter on due diligence and full material disclosure for the finance of technology companies. This relates to the issues of personal liability in this chapter.

Waste of Corporate Assets

Under case law and various state statutes, officers and directors may be individually liable to the shareholders of the corporation for any waste of corporate assets. This liability can arise wherever ownership of assets of the corporation are lost to the public domain or their value is otherwise unnecessarily diminished. The classic defense to waste of corporate assets by officers and directors is the exercise of business judgment rule. That is, officers and directors will not be liable for waste of corporate assets where they lose title to or value in an asset, if they applied their business judgment to prevent such events. In other words, being wrong does not accrue liability as long as reasonable efforts were made with the proper intent. It is inaction and dereliction of duty resulting in damage to the corporate assets that becomes actionable to the officers and directors personally.

This concept becomes germane in light of the slippery and vaporous nature of title to intellectual property. In the U.S., the title to an invention arises in the name of the individual inventor, and remains there until assigned in a writing recorded at the patent office. Furthermore, patents must be applied for in order for patent rights to develop at all. Copyrights for works (including for software) created by consultants also arise in and remain with the consultant, absent a written contract to the contrary. Hence, in the case of patents and copyrights, lack of written contracts with employees, consultants, partners and suppliers may allow the ownership of corporate developments to reside in the individual names of the inventors and authors, not in the corporation's name. Furthermore, publishing, using or selling innovations prior to filing patent applications may result in ownership of the invention entering the public domain.

This kind of inadvertent loss of title to innovations due to inaction can represent a substantial waste of corporate assets where the lost rights are of value. The value may be considerable. (For example, Texas Instruments, a manufacturing company, not an R&D company, now reports that its revenues from royalties from its patents exceed its revenues from manufacturing. Note also that Polaroid recently obtained a patent infringement judgment against Kodak which included a permanent injunction against Kodak from competing in Polaroid's main line of business, and required Kodak to pay Polaroid almost $900,000,000 in cash.) Loss of all or part of a potentially valuable patent portfolio through mere inattention would substantially harm the proper corporate owner.

In this environment, to protect its officers and directors from charges of waste of corporate assets, a corporation need not be patent happy and file patent applications for every hiccup and twist that its engineers may develop. However, a program is needed of conscious application of business judgment to review the company's technological innovations and make explicit judgments as the best course to pursue for patents and copyrights in each case. The corporation may very well best serve its interests by letting a particular innovation lapse into the public domain, and thereby save even the small overhead required to obtain a patent. However, the company should at the least consciously make this decision in pursuit of the best interests of the corporation, in the judgment of the officers and directors. Such a decision to not patent should not simply evolve by default. Otherwise, the loss of title to a valuable patent by simple inattention (and the lack of timely patent applications and proper employee contracts) will leave the officers and directors without the shield of the exercise of business judgment rule.

11

Software and Data Patents: New Developments of 1994

"Patents are not just for gizmos anymore."

Several cases that represent interesting new developments in the U.S. law of intellectual property were reported in 1994. The cases discussed herein focus on computers and medical devices. The computer cases include lessons on how to obtain patents for data structures, how to obtain patents for software, and how to protect data input formats with copyrights. A case is also discussed that further erodes the "business method" barrier to patents. Recent and anticipated amendments to the U.S. patent statute are also discussed.

Data Structure Patents: Two Cases

In the case *In re Lowry*, 32 F.3d 1579, 32 USPQ2d 1031, 48 PTCJ 467 (Fed. Cir. 1994), the patent applicant made a patent application including limitations regarding data structure. Claim 1 was for "a memory for storing data for access by an application program being executed on a data processing system," comprising: a data structure stored in said memory; a single holder data attribute object; a referent attribute data object; and an apex data object stored in said memory. The Patent Office rejected the application saying that the limitations on the data structure were "subject to a mere printed matter rejection and did not qualify for a patent." The Board of Patent Appeals denied the patent saying that Lowry's data structure limitations did not have patentable weight. The Federal Circuit reversed the Board's decisions and said that the data structure does have patentable weight and can merit a patent if it is useful and not obvious. This is an important case that establishes that a data structure in a computer is by itself patentable subject matter. (Note that the claims refer to data in a computer.)

An interesting case to read with Lowry is *In re Warmerdam*, 33 F.3d 1354, 31 USPQ2d 1754 (Fed. Cir. 1994). In this case the court implicitly gave drafting tips for the right and wrong ways to claim data structure in a patent. In claims 1-4, a patent was sought for "a method for generating a data structure," which describes a mathematical algorithm for developing data independent of any computer or means of calculation. These claims were denied as mere manipulation of abstract ideas. Claim 5 describes "a machine having a memory which contains data . . . generated by the method in claims 1-4." This claim was held to be patentable subject matter. Claim 6 claimed "a data structure generated by the method of claims 1-4," and this claim was held to be not patentable subject matter because "data structure" does not imply any hardware or machine. The lesson here is simply to limit the claims for data structure and the method of calculating data structure, to be used in a cited computer or machine, rather than being a free-floating abstract method without citing hardware.

A Software Patent Case

In re Alappat, (Fed. Cir. 1994, cited elsewhere herein) is an interesting case regarding software patents. This case continues the trend to make it easier to get computer software application patents where the application is new and not obvious.

A majority of the 11 judge panel held that the "means plus function" claims should be limited by the implementing structure (in this case a programmed computer) described in the specifications, and hence merit a patent even though the function is a mathematical calculation process performed on data. This "means" should not be read to be over-broad and cover all means (which would include the mental process of thinking through the mathematical steps, which is not patentable). Hence, it is patentable subject matter to claim means plus mathematical function, where the specified means are physical computers or circuits.

Interestingly, the opinion goes on to say that limits on patentable subject matter are narrow and include only "laws of nature, natural phenomenon, and abstract ideas," without mentioning "business methods" (although the opinion later seems to confuse business methods with abstract ideas). Although not at issue here, this may indicate an evolution away from the "business method" patent prohibition.

Alappat gives an important drafting tip for software claims: with means plus function formats for claims, mention hardware in the specification.

In pure method software claims, hardware might best be explicitly included in claim elements in place of the "means for . . ." language to avoid abstract idea objections.

Copyright for Data Input Formats

In *Engineering Dynamics v. Structural Software*, 46 F.3d 408 (5th Cir. 1995), 26 F.3d 1335 (5th Cir. 1994), it was held that the utilitarian function of data input formats for computer programs does not outweigh their expressive purpose so as to preclude copyright protection, since input formats as a whole convey substantial information regarding data which the user needs to gather, and about how the data should be organized in order to run the program properly. That is, the court found that data input formats for computers can benefit from copyright protection, but not necessarily in all cases. The court determined that the extent of copyright protection applicable to the computer user should be approached cautiously since user interfaces are highly functional and functionality is not something generally associated with copyright protection. Likewise, interfaces may contain highly standardized technical information which is also generally not copyrightable, and therefore come near to uncopyrightability. Of course, patent protection is preferable to copyright for computer developments, where it is obtainable. (In *Engineering Dynamics v. Structural Software* (cited above), the product was a structural engineering program applicable to the offshore engineering market.) This case is pending a re-hearing before the Federal Circuit, scheduled for November 4, 1995, and a new decision is expected.

What is a Computer?

An interesting computer case, *In re Paulsen* (Fed. Cir. 1994, cited elsewhere herein) held for a claim containing the term "computer," that "computer" lacks any standard definition but is basically a device that is capable of carrying on calculations. Thus, a prior art reference that discloses a "calculator" meets all the limitations of claims for portable computers, and therefore calculator prior art blocks a patent for subsequent computer art doing similar activities.

Business Method

In re Schrader, 30 USPQ2d 1455 (Fed. Cir. 1994) is an interesting case that indicates explicitly that the old "business method" doctrine may be headed for extinction. The business method doctrine, which has developed in cases over the years, asserts that 35 U.S.C. 101 would prohibit the patenting of a "business method." In *Schrader*, the two-judge majority indicated that claims in the U.S. patent application for a method of competitively bidding on a plurality of related items did not contain statutory subject matter for a patent under Section 101, and therefore, no patent should issue. This was decided on the grounds that (1) a mathematical algorithm was implicit in the claims, even though the claims may imply no more than a step of addition, (2) the claims describe a mathematical problem posed in determining the optimal combination of bids in the auction, and (3) the algorithm was not applied to or limited to physical elements in a device or physical process steps. (The judges found that merely entering bids into records is insufficient to impart physical elements and thereby patentability.) The two judges in the majority were Mayer and Plager.

However, Newman in the dissent, has the better argument. Newman finds statutory subject matter under Section 101. Newman states that the "method involves more than mental steps or theories or plans," citing the *Arrhythmia* case, 958 F.2d, 1053 (Fed. Cir. 1992), 22 USPQ2d 1033 (Fed. Cir. 1992), wherein it was found that a claim must be directed to physical elements or process steps, but wherein a decision regarding medical treatment was found to be such a process step. Newman finds that unpatentable subject matter under Section 101 should be limited to a scientific principle, law of nature, natural phenomenon, abstract idea, or mental steps. Newman then goes on to attack the "method of doing business" grounds for finding non-statutory subject matter under Section 101. Newman finds this "a fuzzy concept" and says, "My guidance is that it be discarded as error-prone, redundant and obsolete." Newman goes on to say, "Any historical distinctions between a method of doing business and a means for carrying it out blur in the complexity of modern business systems," citing *Paine Webber v. Merrill Lynch*, 218 USPQ 212 (D. Del. 1983), regarding the CMA patent, wherein a computerized system of cash management was held, in Newman's view, to be statutory subject matter.

We expect that Newman's dissent is the better law in this case and represents the trend in the cases more that the majority in this case. Our expectation is that the trend in *Arrhythmia*, *Paine Webber v. Merrill Lynch, Alappat*, and the dissent in *Schrader*, will continue. This means that software patents will become even stronger, and will continue to erode old prohibitions against even non-computer applications for business methods (as seen for example, in the Video Jukebox Network patents, such as U.S. Patent No. 5,089,885).

12

The New Paradigm for Software and Financial Products: Stac Electronics

"6/08/94 -- *PERMANENT INJUNCTION [against future infringement] against defendant Microsoft Corp...*
6/13/94 -- *AMENDED JUDGMENT AND ORDER ... Microsoft shall pay Stac $120,000,000.00 ...* "

Civil Docket for Case # 93-CV-413
Stac Electronics v. Microsoft Corp.

A new paradigm for the success of software products has been set. This will be critical for the survival of new products dependent on software (including financial products). It will also be critical for the survival of new companies dependent on those products.

In the *Stac Electronics v. Microsoft Corp.* (cited herein) litigation, resolved in federal court in 1994, a small firm with a software patent (Stac Electronics, which had a patent on the algorithm for its PC hard disc data compression software product) defended its market niche in court against a large infringing corporation (Microsoft).

Apparently, Microsoft had expressed an interest in working with Stac, and did due diligence on the Stac product. In this process, Microsoft determined to copy the compression algorithm of the Stac product. Microsoft then wrote its own code to execute the Stac algorithm and used the code in the Microsoft DOS 6.2 product. (There is also a lesson here in "due diligence" as a form of "industrial espionage.")

Stac sued Microsoft for patent infringement, copyright infringement, and trade secret violations. Stac lost on all counts except the patent infringement count, on which Stac won. Stac received a judgment of about $120 million, plus a permanent injunction against Microsoft to stop further infringement.

After the litigation, for about a week, a lobotomized version of DOS was shipped with the compression feature disabled. DOS manuals were shipped with stickers on the cover warning to ignore the chapter on compression, because the feature had been disabled (pursuant to the federal injunction). After about a week of this, Microsoft cut a deal with Stac by which Microsoft, apparently, paid Stac a large amount of cash, Microsoft made a large capital investment in Stac, and Microsoft received a license to use the algorithm in DOS.

Apparently, after getting caught in court by the patent, Microsoft thought it wanted to deal after all. And Stac found a price that looked good.

If not for its patent and the resulting injunction against Microsoft, Stac would likely be in serious financial trouble today, or out of business. How could Stac sell its product if the same features were available in DOS (or Windows 95) for free? Without a patent, Stac would have just been a free test market project for Microsoft.

Notice that Stac's corporate life was saved by its software patents only (which covered its basic algorithm). Software copyright got nothing for Stac because Microsoft did not copy source code; instead, Microsoft just wrote new code for the same algorithm.

This is the new paradigm. This is how a little software company can save itself from annihilation by a giant with overwhelming market power and financial strength.

This approach also applies to any easily copied product or service that can be protected by software patents. Financial products and services are a prime example of this. New financial products are now receiving patent protection by a species of software patent. This is important because a new financial product can be very quickly copied once it achieves market success.

Immediately after Stac's courtroom victory against Microsoft, the president of Stac announced to the public, correctly we think, that this was the new paradigm for the development and marketing of new software developments. That is, patents must be obtained for new software products, so that they can be protected from easy copying and infringement.

Note that this patent protection for software is superior to the traditional copyright protection. This is because copyright protects little more than copying of source code, and perhaps screen displays and user interfaces, while patents can protect the basic concept of a software product, regardless of the actual source code. In the case of *Stac Electronics v. Microsoft Corp.*, Microsoft avoided a copyright problem but ran afoul of patents.

13

New Developments 1995: Globalization and Software

The first ten months of 1995 brought several important legal developments to U.S. intellectual property law. These changes represent the continuing strengthening of software patents in the United States (and perhaps a corresponding weakening of software copyright), and the further development of U.S. patent law to harmonize with trends in the rest of the world.

The Proposed Guidelines for Software Patent Applications

[In 1996, after this chapter was written, the U.S. Patent Office issued the final guidelines on this subject, replacing the proposed guidelines discussed below. The final guidelines are reprinted as Appendix 3 herein, and are discussed in Chapter 31 hereof. However, the following discussion of the proposed guidelines remains interesting on the points that are unchanged by the final guidelines, and to show the historical development and policy tensions that played out in the development of the final guidelines.]

On June 27, 1995, the U.S. Patent Office published in the Official Gazette proposed "Examination Guidelines for Computer Implemented Inventions" (the so-called "Alappat Guidelines").

These are proposed guidelines to be issued to Patent Office examiners for the review of software oriented patent applications. These guidelines were issued in response to the capitulation of the Patent Office to the Federal Circuit and the pro-software patent trends manifested *In re Alappat* (discussed elsewhere herein).

These guidelines represent a major new pro-patent step forward by the Patent Office. These guidelines will go a long way to resolve the confusion at the Patent Office and in the patent profession as to the best way to format claims for software applications. This in turn will accelerate, standardize, and make easier for applicants, the entire software patent application procedure. In the end, this will make software patents more common, and more valuable.

Perhaps the biggest step forward in the Alappat guidelines is implicit. The guidelines can be taken basically as an instruction from the Commissioner to the examiners to cooperate and facilitate in the examination of software patents.

The Alappat guidelines explicitly direct examiners to deal with software patent applications (so-called "computer implemented inventions") in a "prompt yet complete examination." The guidelines also direct examiners to not limit themselves to just format objections to claims in the initial office action, but also to deal with substantive patentability issues. This alleviates a problem of a lot of software prosecution lately at the U.S. Patent Office, where initial office actions concern themselves mainly with the format of the software claims, and do not deal with the substance of patentability. These "format only" actions led to slow prosecution of software applications. Software patent prosecution was additionally confused because of the lack of agreement among examiners, and lack of agreement between the Patent Office and the Federal Circuit, over the proper form of software claims. Now, under these guidelines, examiners would be directed to bring up format issues in the indicated uniform manner in their first action, and also to deal with substantive patentability problems.

Furthermore, the guidelines direct examiners to "indicate how rejections may be overcome and problems resolved," so that patents may be issued more readily. That is, examiners should indicate where possible, in each official response, how to proceed to obtain a patent. The role of the Patent Office is, then, not to say "no," but to help get to "yes."

Software Claim Formats

The guidelines also go a long way to clarify the proper possible formats for software claims. To be in proper format, a claim must on its face describe a technology that is eligible for a patent under the patent statute. For example, a "statutory machine" is eligible, and a claim in that format may be considered. Since "software" is not mentioned in the statute, the game for software claims must be to format the claims as one of the statutory technologies. Three basic formats are presented:

1. *Software Machines*. The guidelines indicate that "a computer or other programmable apparatus whose actions are directed by a computer program or other form of software is a statutory machine." Hence, a software claim may be in proper format if it reads something like:

> a programmable apparatus for [doing a function] comprising:
>
> (1) means for [doing a function element]
>
> (2) etc.

Apparently, this would also be adequately claimed with claims in the format of:

> a programmable device whose actions are directed by software executing a process comprising:
>
> (1) doing [step 1]
>
> (2) etc.

This latter formulation might also avoid the sort of "Donaldson" problems of interpreting "means plus function" claims that are described later herein.

2. *Software Processes.* A process format for software claims may also work; that is, a "series of specific operational steps to be performed on or with the aid of a computer is a statutory process." Presumably a claim in the following format would fall in this category:

> A process to be performed on or with the aid of a computer, comprising the following steps:
>
> (1) doing [step 1]
>
> (2) etc.

These guidelines clarify classic arguments over process and apparatus formats for claims for software inventions. In the past a lot of confusion has arisen regarding software claims formatted as apparatus claims and/or as process claims. Now these guidelines tell us that both formats work well, and instructs us in how each format should be structured.

The preceding developments in the guidelines are an excellent contribution to the field and will do a lot to expedite and make more consistent software applications within the Patent Office.

3. *A Great New Format: Software in Memory.* A third format presented in the guidelines was somewhat of a surprise. The new format is a great innovation and will be a tremendous help to the profession. The third format offered by the guidelines states, "A computer readable memory [containing software] that can be used to direct a computer to function in a particular manner when used by the computer is a statutory article of manufacture." The guidelines elsewhere go on to state that "a computer readable memory encoded with data representing a computer program that can cause a computer to function in a particular manner or a component of a computer that can be reconfigured with a computer program to operate in a particular fashion can serve as the specific structure corresponding to a means element." That is, "computer program related elements of a computer implemented invention may serve as the specific structure, material or act that correspond to an element of an invention defined by a means plus function limitation."

Therefore, the following claim format should be a statutory article of manufacture:

> "A computer readable memory, encoded with data representing a computer program, that can be used to direct a computer when used by the computer, comprising:
>
> (1) Means for [a function]
>
> (2) etc.

The guidelines would make software claims in this new format much easier to enforce. In the case where patentable software is pirated and sold by a middle man pirate, to legitimate customers of the patent holder, the pirate himself may not be a direct infringer of software claims in the classic formats (that is, claims that are limited to a computer programmed with the software, or the process as executed on a computer by the software). That is, the pirate may have neither the hardware platform to execute the software nor ever actually execute the software, but merely have the equipment to copy the software and the copies of the software on discs. Hence, to pursue an infringement claim against the true culprit, the pirate, a patent holder would have to presumably sue its ultimate legitimate client, the buyer, for infringement and go against the pirate for contributory infringement or induced infringement.

However in the new format, with claims for the software on a disc, the patent holder could sue the middle man pirate only as a direct infringer.

Data Structure Patents

The guidelines go on to indicate that data structure can be patented and how this may be done.

The guidelines would simply state that:

> "a claim that defines an invention as any of the following subject matter should be classified as non-statutory [that is, unpatentable] . . . [for example] a data structure independent of any physical element (i.e., not as implemented on a physical component of a computer such as a computer readable memory to render that component capable of causing a computer to operate in a particular manner. . .)"

Therefore, the direction is that a data structure with any physical element (for example, as implemented on a physical component of a computer such as a computer readable memory to render that component capable of causing a computer to operate in a particular manner) would be statutory, that is, patentable. Hence, a data structure claim of the following format should be a patentable article of manufacture, assuming that it meets the other requirements of being new, useful, and not obvious:

> a computer readable memory containing data with a structure capable of causing a computer to operate in a particular manner, the structure comprising:
>
> Element 1.
>
> Element 2.
>
> etc.

Other comments in the Patent Office legal analysis of the Alappat guidelines, published 3 October 1995, indicate that computer data structure can be claimed by the process of structuring that data in a computer. Also, it is indicated that particular data loaded into a computer for a particular function can also constitute a patentable machine.

The guidelines further state that "an applicant may not use computer program code, in either source or object format, to define the meets and bounds of a claim." Of course, source code or flowcharts or object diagrams or other disclosure could be used, and may be required in some instances, in the specification as support for the claims.

It is further interesting to note, the comment and the guidelines that: "articles of manufacture accomplished by this definition [regarding computer readable memory] consist of two elements:

1. A computer readable storage medium such as a memory device, a compact disc or a floppy disk, and

2. The specific physical configuration of the substrate of the computer readable storage medium that represents data (for example, a computer program) where the storage medium so configured causes a computer to operate in a specific and predetermined manner."

The composite of the two elements is "a storage medium with a particular physical structure and function (for example, one that will impart the functionality represented by the data onto the computer)."

Note that the suggestion here is that data imparts functionality as much as does source code, so that data on a disk should be as patentable as source code on a disk.

Furthermore, note the interesting drafting tip that "for example, a claim that is cast as a 'computer program' but which then recites specific steps to be implemented on or using a computer, should be classified as a 'process'. A claim to simply a 'computer program' that does not define the invention in terms of specific patent steps to be performed on or using a computer should <u>not</u> be classified as a statutory process."

Also note the interesting comment that "the specific words or symbols that constitute a computer program represent the expression of the computer program and as such are a literary creation." Hence, they would not be protectable by patent, but would be protectable by copyright, to the extent the copyright protects at all.

It is further interesting to note that the *In re Alappat* guidelines do nothing to stop the slow death of the business method doctrine. Note that when the guidelines refer to "non-statutory subject matter (that is, abstract ideas, laws of nature and natural phenomena," this list conspicuously does not include "business methods." The implication is that the Patent Office is taking no position to support the idea that business methods are non-statutory subject matter. This is important because this common law idea is an attack used against software patents. As suggested by the dissent in *In re Schrader* (discussed elsewhere herein), the Federal Circuit may eventually directly eliminate this doctrine as an attack on software patents, and on any other type of patent.

In many applications, it would be anticipated that claims for basically the same invention would be each stated in all three formats. Of those three clone claims, the software in memory might be the broadest and most valuable.

Of course, all these claim formats would require adequate support and description in the specification, regarding enablement and best mode, and they would have to be new, useful and not obvious. These remain requirements of all patents.

Easier Allowance of Software Claims: Reconciliation of the Patent Office and the Federal Circuit

It is also quite interesting that the guidelines further state as directions to examiners to "construe any element defined in means plus function language to encompass all reasonable equivalence to the specific structure, material or acts disclosed in a specification corresponding to that means element." That is, means plus functions claims are to be construed by the Patent Office so as to be limited by the specific disclosure in the specification, and are not to be construed in the much broader fashion to include any possible means executing that function.

This is consistent with 35 U.S.C. 112(6), but following this policy is a new reversal of prior policy at the Patent Office. Previously, the Patent Office had caused considerable confusion by insisting on interpreting claims for examination purposes in means plus function format to include all possible means, and not to be limited by the specification as required by statute. This would make claims appear much broader than necessary and has made it quite difficult for certain software claims to be allowed. This policy had been pushed by the Patent Office contrary to clear case holdings by the Federal Circuit. This was a fundamental policy split between the Patent Office and the Federal Circuit and caused considerable confusion for software patents.

However, this explicit Patent Office policy to follow the statute and the Federal Circuit cases on this point, will make it much easier to get means plus function format software claims allowed, if they are properly written and supportive specifications are included.

This policy in the guidelines is a reference to the policy adopted in more detail in the so called *In re Donaldson* examination guidelines, published May 17, 1994, in the Official Gazette of the U.S. Patent Office. These guidelines were written in response to *In re Donaldson*, 29 USPQ2d 1845 (Fed. Cir. 1994).

The *In re Alappat* guidelines together with the *In re Donaldson* guidelines, represent a tremendous step forward by the Patent Office to adopt a pro-software patent stance and to end its struggle against the earlier leadership of the Federal Circuit in this regard. This final pro-patent consensus between the U.S. Patent

Office and the Federal Circuit establishes a uniform national policy in the United States that is pro-software patent and represents a considerable legal and institutional clarification and unification of the law. With the *In re Alappat* guidelines, it is now clear that the split between the Patent Office and the Federal Circuit on this point is over and practice in the area should be tremendously facilitated.

Legal Analysis by PTO

On October 3, 1995, the Patent Office issued a legal analysis supporting the proposed examination guidelines for computer implemented inventions. This was an effort by the Patent Office to provide extensive legal support for its proposed guidelines.

Lobbying the EPO and the JPO

It is also interesting to note that the Patent Office is undertaking a lobbying and educational program with the European Patent Office ("EPO") and the Japanese Patent Office ("JPO"). The intent of this effort is to get the EPO and the JPO to adopt guidelines regarding software applications that are similar or the same as the Alappat guidelines. This would provide for a uniform and faster review of software applications in the three major patent offices in the world.

The Cases Leading to the Guidelines

After *In re Alappat*, 33 F.3d 1526 (Fed. Cir. 1994) was handed down, in which it became clear that software running on a general purpose computer was patentable subject matter, several cases followed. *In re Warmerdam*, 33 F.3d 1354 (Fed. Cir. 1994), was decided by a three judge panel and was basically consistent with *Alappat*. Then *In re Lowery*, 32 F.3d 1579 (Fed.

Cir. 1994) was also decided by a three judge panel in a manner consistent with *Alappat* regarding software patents. However, it is almost amusing to note a later case, *In re Trovato*, 42 F.3d 1376 (Fed. Cir. 1994), which was decided by a three judge panel, in which the majority on the panel were in the minority in *Alappat*. The decision in *Trovato* in effect attempted to overrule *Alappat* and state that software was basically not patentable, although many points of the decision were blatantly wrong on the facts. This case is probably bad law and may be reversed in an *en banc* rehearing. It is also interesting to note that *In re Beauregard*, 53 F.3d 1583 (Fed. Cir. 1994), involved a software patent application which was rejected by the Patent Office. The patent applicant appealed. However, prior to a decision on the case, the Patent Office reversed its policy on software patents, issued the proposed *In re Alappat* guidelines, and effectively dropped its opposition to the *Beauregard* application.

This trend represents the triumph of the software patent, and the resolution of the conflict between the Federal Circuit and the Patent Office in this regard. The Federal Circuit did a tremendous service in properly interpreting U.S. patent statutes regarding the new software technology and in leading the development of the case law in this matter. The Patent Office and particularly the new Patent Commissioner did a great service in finally adopting the wisdom of the majority of the Federal Circuit on this issue. With the new apparent consensus between the Patent Office and the Federal Circuit on the patentability of software, the last major conflict within the development of U.S. patent law regarding software patents has been resolved. We can now see entering a permanent Golden Age of software patents in the United States, typified by faster and cheaper and more predictable application and enforcement of software patents.

The final step to improve the situation would be to amend the patent statute to provide for the publication of pending patent applications and third party opposition to applications. This would solve the problem of the objective difficulty in doing prior art patentability searches for software. This would lend added teeth to the statutory presumption in the United States that issued U.S. patents are valid and, in the end, this would help the industry. The current U.S. Patent Examiner encourages publishing U.S. patent applications and in providing some sort of third party opposition procedure, but it remains to be seen if this change will be made. This would require and amendment to the U.S. patent statute by Congress. The proposals for these amendments are before Congress and we will have to watch developments to see if they are passed.

Software Copyright Shrinks

In this light, it is interesting to note the apparent further shrinking of copyright law to protect software.

A trend has developed in most of the Federal Circuits that copyright could protect the so-called look and feel of software, or the command structure and hierarchy of software. However, in 1995, in the case *Lotus v. Borland*, 49 F.3d 807 (1st Cir. 1995), it was held that the Lotus 1,2,3 command structure could not be protected by U.S. copyright law because it was not copyrightable subject matter. The U.S. Supreme Court recently decided to hear the appeal of this case. Hopefully, the eventual U.S. Supreme Court decision on *Lotus v. Borland* will resolve the question of the copyrightability of software, particular its look and feel and command structure, and resolve the split in the various circuit courts in this regard.

This author feels that contrary to the majority of the circuits, the First Circuit is correct. Copyrights should correctly be applied to actual copying of source code, object code, and images generated and presented on computer monitors or other medium by the software, or other visual works generated and presented by the software. Conventional copyright analysis does not require to be changed to continue this protection. However, this author feels that the recent stretching of copyright law to protect the look and feel of software and command structure was a well meaning but erroneous attempt to stretch copyright law to make it more patent-like in its protection of software.

This expansion of copyright by judges took place before the question of the patentability of software was resolved by the courts and the Patent Office. However, ultimately this expansion of copyright should be seen as incorrect and, indeed, unnecessary in light of the clarification of U.S. patent law regarding software.

Patent protection of software is more difficult to obtain than copyright protection, but patent protection is more profound and generalized, and is independent of specific source code, object code, or visual works generated and presented by the software. Copyright protection is easier and more automatic to obtain, but is less effective and easier to go around.

14

Patents for New Telecommunication Services

"Discovery may be a matter of looking at the old with new eyes."

-Marcel Proust

The telecommunications market is being hit by the leading edge of a tsunami, a developing tidal wave of new telecom services and businesses that will re-orient the entire telecom industry. The battle for dominant market share in these new telecom services will be greatly influenced by a new weapon. The new weapon is the patent for telecom services, which is a new type of patent made possible by recent legal developments.

Telecom service companies that fail to enter the patent wars over new telecom services will be able to continue to provide traditional telecom services in traditional ways. However, without their own telecom service patent portfolios, they may be shut out of some of the most lucrative new telecom services of the future, or face higher operating costs due to the necessity of paying patent royalties to others. New services that already are being staked out with service patents include home shopping, home banking, electronic stores, remote gambling, on-line data services, remote monitoring and control, a variety of interactive services, and various 900 number and automated fax services. Major players already in the game include Citicorp, Microsoft, Merrill Lynch, Prodigy and American Express, and a variety of smaller telecom companies.

This new trend toward telecom service patents means that telecommunication service providers, such as wireless telephone and cable television companies, must now become as interested in patents for new services as telecom equipment companies traditionally have been for their new hardware. This will cause a change in corporate culture for telecom service companies away from open full disclosure cultures and toward confidentiality, patenting, licensing and enforcement. This new culture will be further promoted by the growing generalization of competition across traditional geographic and service-mix boundaries and by the ongoing breakdown of local monopolies and duopolies of traditional telecom companies offering limited services in limited areas. This is a dramatic change in corporate culture and business methods that is just beginning in telecom, but parallels developments in other newly patentable industries, such as software.

This race for new telecommunication service patents is reminiscent of the Oklahoma land rush: the gate is up and the race is on to stake out as wide an area of property as possible, intellectual property in this case, to permit the future development of that property by profitable commercial enterprise. And the new breed of telecom service patents are the deeds to these new intellectual property domains.

The development of a new telecommunication services and businesses is being facilitated by technical development within the current atmosphere of deregulation and privatization. The broad range of technical developments that facilitates this includes wireless communication of voice, data, and video (using a variety of technologies including microwave, cellular, frequency hopping, and direct satellite communication to small ground antennas), cheap smart PCs, fax and fax emulator boards, automatic call response, voice recognition, image recognition, cheap reliable encryption/decryption algorithms, GPS (global positioning system), smart cards, smart phones (and smart just about everything else), dramatically reduced costs of microprocessors and memory, pay-for-call telephone (900 numbers), portable powerful PCs, note-books, personal digital assistants and other equipment, complete cable ground networks, ATM (asynchronous transfer mode) switches for cable, 10 million bit per second cable modems, online databases, intelligent agents (a software technique), interactive television, automatic pagers, and intelligent pagers.

This technical innovation allows telecommunication service providers to have a much larger variety of off-the-shelf hardware to piece together to provide platforms for telecom services. The results of this is that old telecommunication services, such as two-way voice transmission, can be provided in new ways, and that new services can also be provided, such as remote banking and interactive home-shopping. Likewise, as these new systems provide new services, and old services are provided at lower costs, new telecommunication based businesses are facilitated.

The New Strategy: Acquire, Diversify and Patent

With this increased importance of telecom service patents requires that many telecommunication service businesses that in the past have not been patent conscious and have operated in an open full disclosure environment, must now institute corporate confiden-tiality programs to maintain and develop trade secrets and patentable information in a secure confidential environment. This requires a dramatic change in the corporate culture and business procedures of telecom service providers, and in their relationships with their business associates, including partners, employees, investors, consultants and suppliers.

This technological dynamism takes place in an atmosphere of deregulation begun by the break-up of the Bell companies. Because of this, more telecommunication players are performing joint ventures, new ventures and undertaking merger activity to position themselves with new combinations of technical and service capabilities to provide new services in new locations.

For example, John Malone at TCI (this country's largest cable company), in describing his acquisition strategy for his company, says that he "wants to own 20% of everything." It is as if the major players in the industry are stepping up to the roulette table, and betting on every number. The telecom industry is all up in the air; the winners in 5 or 10 years will comprise new mixes of services and technologies; and no one today knows which services, technologies, and combinations will be the winners. So the race is on to get a stake on as much as possible, and hope you win. And patents for the new services are part of that betting.

This trend toward telecommunications service patents also makes due diligence for patents as important for mergers, acquisitions, and other financial transactions in telecommunications, as due diligence for income statements and balance sheets traditionally has been.

At the same time, U.S. patent law is changing to make it clear that software and pure software inventions are now protectable by patents (both as patentable processes and patentable devices, when manifested as devices programmed pursuant to the software in question). In a world where many of the new telecommunication services and businesses can only be provided as a practical matter by some degree of automatic programmed machinery, patents on the software and programmed machinery effectively monopolize the only means of providing the new services and businesses. Hence, patenting the necessary software applications in effect restrains any competitors from providing the same services or businesses. That is, the key to patenting a service, is to patent the computer application necessary to implement the service.

Furthermore, especially through the route of software patents, the old limitations against patenting business methods may be breaking down. Indeed, we are increasingly seeing patents issued in the United States for what appear to be pure business methods, especially where they entail a degree of new hardware or software to implement themselves. Business method patents are a rapidly changing area of the law and should be monitored closely. Although the final death of the old prohibition on business method patents has probably not yet arrived, the trend in the law is clearly toward the increasing patentability of more and more subject matter.

Patent Law Developments: Software and Services

The grand-daddy of all software/service patent strategies is none other than Merrill Lynch. In the primordial big bang case on the subject, *Paine Webber v. Merrill Lynch* (cited herein), a 1983 federal case (1983 is ancient in this industry's law, just like in this industry's technology), Merrill Lynch sued Paine Webber to enforce Merrill Lynch's very broad patent on the cash management account (U.S. Patent No. 4,346,442 for "Securities Brokerage-Cash Management System"). This is, practically speaking, a patent on a type of brokerage account. The patent was written ostensibly to cover using a computer to keep the books as particularly necessary for this kind of combined brokerage/checking/credit account. Although the patent may be open to attack for lack of novelty, obviousness and, a shrinking group of commentators would say, the business method exclusion, Paine Webber lost its motion for summary judgment on the question of patentable subject matter, and settled with Merrill Lynch. (Indeed, now there is an increasing variety of financial product patents enabled by this case.

Some of these financial product patents are owned by such financial companies as Citibank, Aetna Life Insurance, and a savings and loan association in Princeton, New Jersey that is active in this field. They cover products including new types of mutual funds, insurance products, and annuity-like savings plans. As discussed below, Citicorp is also now litigating to enforce its patents for home banking.) But what this case meant on a larger level, is that a new service could be captured by patenting computer applications essential to that service. The service in this case happened to be financial, but the principle applies exactly to telecom services or any other computer facilitated services.

A 1989 Federal Circuit case entitled *In re Iwahashi* (cited herein), made it clearer that software is patentable subject matter in the United States. The patent in this suit covered a computer program that calculated a coefficient pursuant to a mathematical formula and then stored the answer in memory. This is the type of calculator program that might have been the most susceptible to attack as merely preempting a mathematical formula. However, the court saw this subject matter as no problem to patentability, reasoning that a mathematical formula was not preempted by the patent in question. Instead, the court decided that the invention constituted a device, being a computer, that was programmed in this particular manner. The patent was upheld. The judgment was based on patent claims that included a physical element of computer memory and the step of storing the calculation results in that memory.

A later case, *Arrhythmia Research Technology v. Corazonix Corporation*, 22 USPQ2d 1033 (Fed. Cir. 1992), enforced U.S. Patent No. 4,422,459 that was granted for "Electrocardiographic Means and Method for Detecting Potential Ventricular Tachycardia," to help predict the likelihood of future heart attacks. The "method" in the patent was an algorithm that translates an EKG signal into a quantified assessment of the patient's cardiac function. The appellate court's decision overturned the decision of the lower court that invalidated the patent protection for the mathematical algorithm, where the lower court incorrectly reasoned that software patent algorithms were not acceptable subject matter for patent protection. In reversing the lower court, the circuit court held that the previous case law that might have excluded software algorithm patents was limited and did not restrain the patentability of software or algorithms used in software. This ruling is part of a general trend by the Federal Circuit to establish the patentability of software. In a more recent case, *In re Alappat* (Fed. Cir. 1994, cited herein) the same court allowed a patent for a computer algorithm for image processing by a computer.

The legal developments to permit software patents, and by this avenue various service and financial product patents, have been met with considerable controversy. However, this legal development is the result of applying traditional patent law principles (the substance of the patent statute has not been amended to make this happen) to new technologies as they emerge. The resulting case law has been steadily developing in the same direction for over a decade. That is, software/service patents are here to stay and will be enforced. The controversy about this development is just part of the education and adaptation process, and it is now time for all successful businesses to learn about and adapt to this new approach.

The Latest Fights to Enforce Software Patents

We have recently seen these software patents successfully enforced in court, for example, in the *Stac Electronics v. Microsoft Corp.* (cited herein) litigation. In this litigation, resolved in 1994, a small firm with a valid software patent (Stac Electronics, which had a patent on the algorithm for its PC hard disc data compression software product) defended its market niche in court against a large infringing corporation (Microsoft). Immediately after Stac's courtroom victory against Microsoft, the president of Stac announced to the public, correctly we think, that this was the new paradigm for the development and marketing of new software developments. That is, patents must be obtained for new software products, so that they can be protected from easy copying and infringement. This is directly applicable to new telecom services, which can usually be readily reverse engineered, once they are first marketed.

Note that this patent protection for software is superior to the traditional copyright protection. This is because copyright protects little more than copying of source code, and perhaps screen displays and user interfaces, while patents can protect the basic concept of a software product, regardless of the actual source code. In the case of *Stac Electronics v. Microsoft Corp.*, apparently no source code was copied; however, Microsoft reprogrammed its own source code using the same patented algorithms as Stac. Hence, Microsoft avoided a copyright problem but ran afoul of patents.

The recent issue of the Compton's multimedia patent plays out these patent developments in the communications and information industry. Compton's received a very broad U.S. patent for multimedia products not long ago, and they then announced broad plans to enforce and collect royalties on the patent from the entire multimedia industry. An uproar resulted, based on the common complaint that the patent was incorrectly issued, since it did not represent any novelty at the time of application. This led the Patent Office to review the patent in light of the extensive earlier use of the same ideas by others, evidence of which Compton's competitors in the industry were happy to provide. As a result, the Compton's patent was eventually withdrawn by the Patent Office.

This is an example of an ultimately unsuccessful attempt by Compton's to use aggressive patent strategies to obtain a dramatic beachhead in a new information service market. Compton's was ultimately unsuccessful, whereas Stac was apparently completely successful. The conclusion on these two recent cases is that we will see more of this patent strategy in a broader range of telecommunications services, but the results of the strategy are not guaranteed in any particular case, and hinge on the specific facts of the situation. In any case, all the players in the telecom industry would be well advised to begin to arm themselves with their own patent portfolios, even if only for defensive purposes.

In a very recent development along these lines, in November 1994, Citicorp sued Online Resources, alleging infringement by Online of three Citicorp patents for home banking by phone. This litigation is in a very early stage and it is too soon to know how it will develop, but Online has counterclaimed against Citicorp to invalidate the Citicorp patents. It is quite interesting that Online has broad patents of its own for home banking (see below). Also, Online has beaten Citicorp to the home banking market in the Washington, D.C. area in a deal with NationsBank, and is dramatically undercutting Citicorp's unit price in this market. This litigation by Citicorp represents another entry by a major player into the service patent strategy with a telecom nexus. (It is also rumored that one of the major New York banks owns and is quietly licensing a recent patent for a type of mutual fund, which takes a software/service approach.)

The Katz L.P. Interactive Patents

A similar patent enforcement situation is now developing regarding the Katz patents, and directly involves telecom service patents with a software orientation. Ronald A. Katz has formed Ronald A. Katz Technology License, L.P., which is the owner of about 28 issued and about 17 pending U.S. patents for interactive telephone communication technology. Most of these appear to be software service patents with indicated hardware platforms. Some of these patents are quite broad, and the Katz L.P. has announced a policy of demanding licenses from infringers and collecting royalties on these patents.

The Katz L.P.'s first litigation enforcement activity ended in favor of Katz in the case of *West Interactive Corporation v. First Data Resources Inc.*, 23 USPQ2d 1927 (Fed. Cir. 1992) a case brought in 1989, and settled in 1992. This resulted reportedly in a settlement of $4.4 million paid to Katz in settlement of past patent infringement, with ongoing royalties.

Katz received his original interactive telephone service patents and sold them to First Data Resources ("FDR") in 1988, then a unit of American Express Information Services. After Amex spun off FDR to the public in 1992, Katz bought the patents back. Hence, the Katz L.P. situation represents the results of a step by another major corporation into telecommunication service patents. We can now expect this to be a trend, and major players must expect to get into the game with their own patent portfolios, or lose the patent race by default.

Since the *West Interactive* settlement, it is reported that licensees of Katz L.P. now include MCI, American Express, Home Shopping Network, Sprint, and the TCI cable company. Mr. Katz appears to have become a multi-millionaire by royalties alone.

However, AT&T has refused to pay royalties to Katz, and in July 1997, AT&T was sued by MCI (a Katz licensee) for infringement of the Katz patents. AT&T may be the first Katz defendant with the will and the cash to attempt to invalidate the Katz portfolio, if possible. But it appears that the Katz L.P. will be able to ride to this legal battle with AT&T on the back of MCI's checkbook.

It is unclear how the Katz situation will play out. There are some experienced people in the interactive telephone industry who feel that the Katz patents will suffer the same fate as Compton's. That is, they feel that at least some of the patent applications clearly were not new at the time that they were applied for and that there were a variety of other people who had the same services available on the market in the public domain years prior to the Katz applications. On the other hand, the Katz patent portfolio is large and has many independent claims. A license on reasonable terms may make more business sense than a broad legal attack on the Katz portfolio, even if the entire portfolio were questionable.

In any case, this path is clearly laid out for more major players, along with American Express, MCI, Compton's and Citicorp, to undertake this software patent strategy for communication services.

The 1995 Survey

In 1995, we undertook an informal survey of who was getting what out of the Patent Office, particularly in areas related to services through pay-per-call telephony and cable television.

We find a new wave of patents for cable services, although to date this type of patent has been pursued primarily by smaller more innovative companies. We can anticipate that major players in telecommunications service will soon get into telecommunication service patents, and may have pending unpublished patent applications of this type, but they have yet to appear in the form of issued patents in the Patent Office.

For example, U.S. patents by smaller players in the 900 pay-per-call industry, include No. 4,893,333 for "interactive facsimile system and method of information retrieval"; No. 4,974,254 for "interactive data retrieval system for producing facsimile reports"; No. 4,918,722 for "control of electronic information delivery; and No. 4,941,170 for "facsimile transmission systems." These patents use various kinds of 900 numbers to call in to service providers and generate automatic fax-back of certain types of fax reports. These patents each attempt to patent not a new hardware "box," or even really a new software system. Practically speaking, what they attempt to do is capture a new type of business or service incorporating databases, 900 numbers and fax-back delivery. They do this by using the language of computer and microprocessor applications.

Directly in the field of new cable services, we have recently seen the development of the family of video jukebox patents. These are owned by the Video Jukebox Network, Inc., and include U.S. Patent No. 5,089,885 for "Telephone Access Display System with Remote Monitoring." These are very interesting patents because they directly pursue patents on the service provided through the video jukebox cable channel. These patents provide for no new hardware, no new combinations of old hardware, and, indeed, they do not even provide for new software. Instead, these patents directly pursue what we would call the business method of: projecting a menu on a cable channel of code numbers for rock video titles, calling in to a central station on a 900 number and indicating the serial number preferred, and then broadcasting the selected rock video on the common view cable channel. There is nothing novel about these patents except the general method itself, which uses other pre-existing or proprietary technology in a new way. These patents have yet to be litigated and may be open to attack, but on their face they seem at least to be reasonably defensible. They do represent an attempt to own new cable services without bothering with the claim language of computer applications, and in this sense open exciting new ground.

Our recent survey of the major cable service providers indicates that the major providers have yet to advance to the stage of issued U.S. patents or pending foreign patents in telecommunication services. (Pending U.S. patents are unpublished and secret at this time, although there is a proposal to change the U.S. patent law to publish pending U.S. patents prior to issuance, as is done in the rest of the world.) With smaller players already active in telecommunication service patents, with a clear development of the law encouraging this, and with some major players in other areas, such as American Express, Citicorp, and Compton's, already obtaining wide publicity for their service patents, we can now expect that the major players in all telecommunications areas must eventually move actively into this area.

We note that Time Warner (a major cable company) has only very recently become active in the U.S. patent arena, but with a hardware orientation. Likewise, Viacom recently has become active with foreign patent applications, and we can anticipate that both Viacom and Time Warner might have further U.S. patents pending that are yet to be publicly visible.

GTE, of course, is active, but more on the hardware end. Scientific Atlanta is active with hardware and software, but has yet to obtain telecom service patents.

The Surprise Leaders

Our survey indicates who may be the surprising current leaders in telecom service patents, and they are not telecom companies. It is very interesting that Prodigy, the online data service, has become quite active in home shopping patents, Online Resources, Inc. has become active in patents regarding online retail banking and home shopping, and Citicorp is also active in home banking patents. Likewise, a company called Teleaction Corp. has become active in an interesting area regarding electronic stores. It seems that the development of corporate activity in telecom service patents is following the legal development of the patentability of services. That is, service patentability arose out of software patentability (particularly for financial products), and now software, information and financial companies may be beating the traditional telecom companies to the new telecom service patents. (On this point, experience in the new patent law seems to have been worth more than experience in the telecom business.)

Also note the recent deal with Microsoft and TCI, to put Microsoft's pending on-line data service on TCI's cable system. Microsoft recently committed to aggressively pursuing and enforcing software patents (even before the Stac episode). So now that Microsoft is about to become a telecom service company, and is in bed with TCI, we can expect at least one more big player to hit the telecom service patent field soon.

There is lots of software patent activity by smaller players in the area of home shopping, pay-per-view, video on demand, and interactive television. There is also some activity for gambling and lottery concepts, many with telecommunication attributes.

The New Strategy

The strategic business conclusions from all this is that whenever a new line of business is contemplated in telecommunication services, an integrated intellectual property strategy should be contemplated to protect that service or business, including patents, copyrights, trademarks and trade secrets. This can be done to maximize market potential, to increase unit sales, to increase unit price, to maintain profit margins, to suppress competition, and to promote brand differentiation.

The players should also note that intellectual property ownership for new telecommunication services and businesses can be quite easy to lose, particularly prior to the application for patents and before copyright registration. An integrated intellectual property strategy should be initiated early in the business development process and include protection of ownership. (Note that a working prototype is not necessary for a patent application.) Effort should also be taken to avoid developing a product that may infringe the patent rights of others, since infringement by corporations may lead to civil and criminal liability to the corporation and its officers and directors. The general rule is that if it is time to spend money on development, then it is probably time to consider an immediate patent application, since the conceptual stage has been reached that is necessary for patenting, and that is adequate for theft of the idea by competitors.

Prophylactic routines should be established as part of the corporate culture to maintain a general attitude of confidentiality and protection of current and future intellectual property. Among other things, this requires appropriate contractual documentation with all employees, consultants, partners, suppliers, potential investors, and business associates. In the context of doing due diligence for a prospective business transaction, be aware that where the deal does not close the due diligence may turn into de facto industrial espionage. This corporate confidentiality program would include nondisclosure, nonuse, non-competition and assignments that are adequate for all future developments, particularly from employees, consultants and vendors. In particular, note that title to intellectual property is often defective and take steps to identify and cure title deflects. Maintain internal audit procedures to avoid the unauthorized use on corporate equipment of unlicensed software of others. Also note that regarding inadvertent infringement, a written opinion of non-infringement by outside patent counsel may serve to minimize personal criminal liability of officers and directors, and to reduce exposure for damages to the corporation by as much as two-thirds, in the event that eventual inadvertent infringement does occur.

The Patent Shoot-Out for Market Dominance

It is traditional in the U.S. that the ultimate winner in the race to capture the lead position in a new industry is often determined by the outcome of a patent shoot-out.

To name just a few of many examples, it happened with cars when Henry Ford was sued by the holder of the patent for the automobile. It happened with the airplane when the Wright Brothers went to court with Glen Curtis. It happened with the radio when Thomas Edison went to court with Marconni. More recently, it happened with self-developing film in *Polaroid Corp. v. Eastman Kodak Co* (cited herein). (Polaroid won. Kodak is out of the business, and almost $1 billion poorer.) It happened with Amgen with the new drug EPO.

With the development of software patents and the erosion of the "business method" doctrine, it happened more recently with software patents in *Paine Webber v. Merrill Lynch*, cited herein (Merrill won), *Arrhythmia v. Corazonix*, cited herein (Arrhythmia won), and *Stac Electronics v. Microsoft Corp.*, cited herein (Microsoft lost, so they bought the winner).

Note that in all the cases just cited (except for the two oldest, about the car and the airplane), the patent holder won in court, and then in the marketplace.

We can expect the same thing to happen with the current development of new telecommunication services and businesses. The pattern will be along classic lines (except that the patents will be for services): first a rush to market with a new service. Then a number of leaders will establish themselves with profitable market shares. Then, the leaders will have a patent fight. The fight may be joined, or even won, by a small player, or even an individual with a superior patent. The ultimate winner of the patent fight will take the dominate position in the industry. (If the winner of the patent fight is a smaller player or an individual, they will sell out to a big player.)

To date, smaller start-up companies have been the nimblest in the race to the Patent Office to stake-out patent ownership to new telecom services and businesses. But the first signs are developing that the major telecom players are also entering the field, and we can expect the contest to develop quickly from here.

15

Patents for Financial Services and Program Trading Strategies

New legal developments make it clear that software-enabled financial services and algorithms for program trading strategies are now patentable. Indeed, still breaking legal developments clarify for the first time that any software, whether for financial services, financial products, or otherwise, is candidate subject matter for a patent. If the software or algorithm meets the relatively low legal hurdles for being new, useful and not obvious, then U.S. patents are obtainable if pursued early and diligently. These patents are broadly construed to exclude others from using the same general ideas claimed in the patents, or their equivalents, and can be obtained without publicly disclosing the source code for the software. Although overlapping copyright protection can be simultaneously obtained for software, patent protection is stronger, when available. These software patents are being newly enforced, and are proving of great value.

Software Patents for Services

One impact of the patentability of software and software algorithms is that financial products and services (which may be provided on a profitable basis to a mass market only by computers) can be captured by patenting the software to deliver the product or service. For example, a new type of brokerage account might not be patentable subject matter; however, the software to keep the books and records for the brokerage account may indeed be patentable. Therefore, as a practical matter, a patent on the record keeping software for the new account would preempt all offerings of that brokerage account. Anyone could, without a license for the patent, offer the brokerage account, but they would have to keep the books and records manually.

How It Started

This patent approach for financial services is relatively new and was first tried when Merrill Lynch obtained a patent for its CMA account in 1982. See, U.S. Patent No. 4,346,442 for a "Securities Brokerage-Cash Management System." Merrill Lynch then sought to enforce this patent against Paine Webber, when Paine Webber attempted to copy the basic idea of the CMA account, using computers of course to keep the books and records. *See, Paine Webber v. Merrill Lynch*, 564 F. Supp. 1358 (D. Del. 1983). In this case, Paine Webber moved for summary judgment that the patent was invalid because it covered non-patentable subject matter, being in effect a software patent. The court refused to invalidate the CMA patent in this manner, and ruled that mere software applications may be patentable. Paine Webber settled the case, and today the CMA account patent still stands. Although the CMA patent claims, on its face, merely a computer system that keeps the books and records for the CMA-type account, the patent in effect preempts the CMA account for practical purposes, since it is not economic to provide the service without computerized records. Manual records, as is the case for most financial products and services, are simply impractical.

Early Legal Confusion

When software patents were first pursued in the 1970's, the earliest patent cases regarding software were confused, and often bordered on being incorrect. The fundamental problem was that mental processes and laws of nature, including mathematical laws, are not patentable subject matter, and federal judges thought that computers were little more than big calculators. The misconception was that anything computers could do was merely a mental process in disguise, and hence was of suspect patentability. This legal confusion is apparent, for example, in a 1972 U.S. Supreme Court case, *Gottschalk v. Benson* (cited herein). In that case, the Supreme Court took the view that a computer is little more than a complicated calculator and ruled that a computer process essentially only manipulates numbers. As such, the court reasoned that the computer process in question merely preempted a mathematical algorithm and therefore should not be patentable. Twenty years later, today's laws have evolved to a more correct position, which in effect overrules the most egregious anti-software interpretations of *Gottschalk*.

Likewise, in a 1978 decision, *Park v. Flook* (cited herein), by the U.S. Supreme Court, a pure mathematical formula was denied a patent when no physical activity took place and computers and computerizations were not part of the claims. These two early software decisions led to considerable confusion and uncertainty, and pessimism, about how or even whether software patents could be obtained.

In 1981, things started to get more optimistic when the Supreme Court handed down *Diamond v. Diehr*, 450 U.S. 175, 209 USPQ 1 (1981) and began to move away from the *Gottschalk* holding. In *Diamond*, the Supreme Court acknowledged that computers and software have a function in industrial processes that is unrelated to mental mathematical calculations. Specifically, this case upheld a patent for an industrial process that incorporated computers, computer software, and computer algorithms, even though one of the steps in the process recited a mathematical algorithm. The Supreme Court in this case eroded the old anti-software commandment, "thou shall not patent mathematics," and things became more optimistic for software patents.

A New Court Gets It Right

Then in 1983, the Federal Circuit Court of Appeals was created and given exclusive jurisdiction over appeals for patent litigation. This court was created to be pro-patent and began to uphold and enforce the majority of patents before it. This tremendously strengthened the legal presumption that all patents issued by the U.S. Patent Office are valid. This represented a considerable change: prior to the creation of the Federal Circuit, about 75% of the patents litigated in the federal court were ruled to be invalid. Just prior to this, things improved with the 1982 decision of the Court of Customs of Patent Appeals (the immediate predecessor of the Federal Circuit), in a case called *In re Pardo* (cited herein). In this case, the court ruled that a pure programming technique was patentable, in this case a computer compiler algorithm. The court stated correctly that there is a difference between an unpatentable mathematical algorithm (meaning, in this case, a human mental process) and a patentable computer algorithm (meaning a method, whether mathematical or not, done by a computer).

These optimistic developments continued with *Northern Telecom, Inc. v. Datapoint Corp.*, 908 F.2d 931, 15 USPQ 1321 (Fed. Cir. 1990). This case indicated that a valid patent needed to describe the inventive features of a software patent only at the flow chart level, and that this did not require disclosure of source code. The court even suggested that disclosure of source code was discouraged because of its bulk and the difficulty of the average court in interpreting what source code means.

Things improved further in the Federal Circuit case *In re Iwahashi* (cited herein), in November 1989, which further supported the patentability of software. In this case, a patent was ruled to be valid even though it covered a computer program that only calculated a coefficient pursuant to a mathematical formula, and stored the answer in memory. This type of "calculator program" might have been susceptible to attack as merely pre-empting a mathematical formula. However, the patent was upheld basically because the claims included reference to the physical element of computer memory, and the physical step of storing the calculation results in the memory.

Two Key New Algorithm Cases

A wonderful new case for software algorithms developed recently in 1992. *See Arrhythmia Research Technology Inc. v. Corazonix Corp.*, 22 USPQ2d 1033 (Fed. Cir. 1992). In this case, the Federal Circuit upheld a patent for a software algorithm for analyzing electrocardiographic signals to predict heart attacks. The court held that this is patentable subject matter, overturning a lower court decision to the contrary. Obviously, of course, if a software algorithm to do a statistical analysis of electrocardiograph signals to predict a future heart attack (and to trigger specific drug therapies) is patentable subject matter, then a software algorithm to do a statistical analysis of previous stock market activity in order to predict future stock market activity (and to trigger buy or sell signals) is also patentable subject matter. The only further requirement for patentability for such an algorithm would be that it meets the relatively low legal hurdles being new, useful and not obvious.

In an even newer and more optimistic case for software algorithm patents, *In re Alappat* (Fed. Cir. 1994, cited herein), the entire Federal Circuit upheld a patent for a computer algorithm for image processing by computer. In this case, the court ruled that a general purpose programmable computer running a specific software algorithm is clearly patentable subject matter. (Although the reasoning is convoluted, the court seems quite impressed by the somewhat tortured legal argument that a general purpose computer running a specific program is in effect equivalent to a dedicated special purpose machine hard wired with a specialized circuit to perform the same program and, hence, is fully patentable subject matter as a new type of machine!)

Enforcement of Software Patents: The New Paradigm

In the enforcement end we saw, in 1994, the case of *Stac Electronics v. Microsoft Corp* (cited herein). In this case, little Stac sued giant Microsoft for two counts of software patent infringement. Stac alleged that Microsoft stole the algorithm for data compression and decompression in a Stac software product. The judgment upheld Stac's allegation, and Stac obtained a judgment for about $100 million from Microsoft, and a permanent injunction against Microsoft from further use of the Stac algorithm for the duration of the two patents in question. (As a result of this injunction, Microsoft had to withdraw DOS 6.2 from the market, and issue a "dumb" version of DOS in its place, with the data compression function permanently disabled.) Rather than fight an appeal of this case, Microsoft then cut a settlement with Stac Electronics in which Microsoft paid Stac a large sum of money and further purchased a substantial quantity of the stock of Stac Electronics. In other words, the software giant, Microsoft, concluded that it couldn't beat the little start-up company, Stac Electronics, that was protected by a good patent for its software algorithm. Therefore, Microsoft decided if it couldn't beat them, it would just buy them. This was an option that the owners of

Stac Electronics did not have to extend to Microsoft, but presumably the owners of Stac Electronics decided that they could be quite happy if their price were met.

In another, still breaking, algorithm patent enforcement action, Comsat Corp. filed a patent infringement suit, in April 1995, against General Instruments and two cable companies, alleging infringement of a single Comsat patent for a decryption algorithm for cable television "decoder boxes." Since these decoder boxes are apparently used by a large number of cable customers in the U.S., the eventual value of this one patent, if Comsat wins everything that it is seeking in this case, could be huge. It is impossible to predict at this point how this case will develop, but is it is clear that another major player is betting heavily on the validity and enforcement of algorithm patents.

With these new patent law developments, and the exemplary litigation between Stac and Microsoft, the paradigm for new software developers has been set. If they have a successful new product, then in order to protect themselves against being crushed and ripped off by giant companies, they should get and enforce patent protection.

Examples of the New Patents

Word of these opportunities is slowly leaking into the software industry in general, and into the financial services, financial products, and program trading industries in particular. For example, in addition to the CMA patent, there are a variety of financial product patents now in force in the United States.

Perhaps most interesting for artificial intelligence and program trading applications are patents specifically in these areas, including an "Automated Securities Trading System," U.S. Patent No. 4,674,044, assigned to Merrill Lynch; and an "Automated

Investment System," U.S. Patent No. 4,751,640, assigned to Citibank. Clearly, major players are involved in the new game of patents for program trading rules.

In the context of program trading and related financial applications, where successful ideas travel quickly and are easily copied, and where the personnel that understand these concepts are highly mobile, patent protection can be essential to capture the complete value of new developments implemented by software.

These patentable software applications may include program trading data, analysis packages forecasting future trends, software packages analyzing and graphically presenting financial information, and market simulation packages. Indeed, many of the past revolutionary advances in computer applications for financial services and trading would probably have been patentable if patent applications for them had only been filed when they were new. (This would probably have included broad patents on the basic idea of program trading itself.) In the future, we can expect to see patents on new applications of neural networks to trading, applications of genetic algorithms to trading, search and screening software for large financial databases, Internet applications, systems of profitable trading rules, the development and commercialization of new market indexes, and other new developments.

Important Rules

In this regard, it is important to note that patents must be pursued in a timely manner, or the new software developments will be lost forever to the public domain, without royalty. Specifically, in the United States, when a new patentable development is conceived, a patent application must be on file in the U.S. Patent Office within 12 months of the time that the invention is first sold to the public, offered for sale to the public, described in a printed publication, or otherwise taught the public. Hence, it is important to maintain new patentable software developments as confidential until a U.S. patent application is filed.

It is useful to note that U.S. algorithm patent applications do not have to divulge source code to implement the algorithms. Therefore, a new software applications developer may have his cake and eat it too with software patents; that is, they may be able to, in effect, protect the source code without divulging it.

Also note that a patent protects not only the invention that is claimed, but also all of its legal equivalents. Therefore, a patent should not be avoidable by subsequent competitors who merely make small inconsequential changes to the invention, or who reprogram the source code from scratch without copying a single line of the original source code, if they in effect practice the same general algorithm claimed and described in the patent.

In effect, then, a U.S. patent, which may cost $10,000--$15,000 to draft and file in the U.S. Patent Office, may provide the owners with a legal monopoly for approximately two decades for a new program trading technique or other financial application that may be worth many times that amount, and have cost that many multiples of that amount to develop. For example, what would have been the value of a 17-year monopoly on the basic concept of program trading, if only the original developer of the idea had thought to file a patent application for it?

More Good News Coming

In a startling development that unfolded in 1995, as this chapter was being written, a radical and unexpected pro-software patent case appeared, involving the Federal Circuit appeal of *In re Beauregard* (cited herein). Beauregard filed a patent application directed simply to a specific new software program embodied on a computer memory (such as a floppy disk). No "physical element," such as a computer, was mentioned. This patent application was rejected by the Patent Office as unpatentable subject matter, on the grounds that it was analogous to simple "printed matter." The Patent Office suggested that the idea is the subject only for the far weaker copyright protection. The patent applicant appealed the Patent Office decision to the Federal Circuit. On April 26, 1995, the Patent Office, in a surprise move, filed a motion with the Federal Circuit asking the case be dismissed. In the motion, the Patent Office said that it had reversed its position on this form of pure software patent application. The Patent Office informed the Federal Circuit that it intended to draft regulations for its patent examiners, to be ready perhaps as early as the end of May 1995, instructing them that patent claims for specific new software on memory medium is patentable subject matter, if the software program itself is new, useful, and not obvious.

The tremendous advantage of this lies in how these patents may be enforced and who they may be enforced against. Previously, software patents had to mention specific physical elements such as computers. Therefore, an infringer would be only someone who used the software on the claimed computer element. This would require a software developer company whose ideas had been stolen to sue the ultimate user/customer as the infringer. The user may in fact be an innocent party who simply purchased software stolen by a software pirate, and in any case, it is difficult in the marketplace to sue a customer. However, if the software pirate never actually was caught in the act of using the software on a computer, but merely copied it and sold it on a floppy disk, then the pirate would not be a literal patent infringer (although he may be sued in a more difficult derivative action for contributory patent infringement or induced infringement). Many software companies who have been the victim of piracy find it difficult or unadvisable in the marketplace to sue even one of their clients for infringement in order to get at an intermediate software pirate. However, if the

Patent Office allows claims drafted in the format of software on memory medium such as a floppy disk, then the intermediate software pirate who merely copies software and puts it on a disk for sale would be a direct patent infringer. This may allow direct enforcement against the intermediate software pirate without involving any end-user/customer client in infringement litigation.

[After this chapter was written, the US Patent and Trademark Office issued proposed guidelines for software patents, which are discussed in Chapter 13 hereof. With modifications, the final PTO software patent guidelines were subsequently issued, and are discussed in Chapter 31 hereof, and reprinted in their entirety in Appendix 3 hereof. Also, after this chapter was written, further case law on financial service patents was initiated with the *State Street Bank* case, discussed in Chapter 31 hereof.]

16

Software Patents: Historical Development and the Limits of Copyrights

Today it is easier than ever before to patent software and software/hardware inventions. Many new product developments in a wide variety of industries (including medical devices, telecommunications, and financial products) are either pure software programs or hybrid software/hardware inventions.

For example, in the medical device industry, these developments are found in many areas, including image processing (in magnetic resonance, sonar, and CAT scans); expert diagnostic systems; diagnostic software; instrument control (for example, for laser surgery); specific instrument design and manufacture (such as endoscopic and arthroscopic tools); the incorporation of programmable chips and ROM chips into disposables; and computer-aided design/computer-aided manufacturing applications to design and prototyping (including stereolithography).

In the past, patent law has been ambiguous at best and hostile at worst toward the idea of patenting software, particularly pure software programming inventions. Recent legal and business developments in the world of software patents, however, including those for the medical industry, indicate that pure software patents are available, enforceable, and superior to copyrights for software and hybrid software/hardware inventions.

A random review of recently issued U.S. patents indicates a variety of pure software and medical device/software patents. For example, to list but three of many, in the device/software field we see "Magnetic Resonance Imaging System" (No. 4,983,918); "NMR Spectroscopy with Phase Encoding within a Selected Voxel" (No. 4,983,920); and "Analyzer of Partial Molecular Structures" (No. 4,987,548). Many of these patents merely describe, at the macro-flowchart level, software algorithms used by specific programs for specific computer applications, sometimes in association with known hardware elements. Our informal survey of recent medical software patents also indicates patents for medical billing software algorithms, medical expert systems, and medical user-interactive expert systems.

Cash From Software Patents

Software patents can be cash-producing assets. For instance, Texas Instruments' profits from royalties, much of which are from software or software/hardware patents, now exceed their operating profits as a manufacturing company. In addition, changes in medical equipment software are often critically important to a manufacturer's success or failure in the marketplace. For example, real-time Doppler ultrasound scanning (a significant addition to the field of diagnostic imaging that often enables clinicians to use a safer alternative to angiography) is largely the result of improvements in the software that processes ultrasonic signals. Recent legal developments also show that the federal courts are continuing to enforce most patents and to grant large judgments for infringement. For example, witness the $937 million judgment against Kodak for infringement of Polaroid patents.

In view of these enormous cash flows, many companies that once rejected the idea that software should be patentable (including IBM) are now putting major efforts into the development, patenting, and licensing of software and software/hardware inventions. As computer hardware increasingly becomes a commodity manufacturing business, with commensurate pressure on profit margins, software patenting and licensing will become correspondingly common.

The U.S. patent statutes specify that a new and useful invention may receive a patent if the invention is not obvious, if it has not been sold or publicly disclosed for more than a year before a patent application is made, and if the application is properly drafted and prosecuted. A U.S. patent, once issued, gives the holder a temporary monopoly in this country: it prevents others from making, using, selling, and importing the invention for the term of the patent. Licensees sold a license by the patent holder may use the invention if they satisfy license requirements, including payment of royalties.

Foreign patents are also possible in most developed countries. Note, however, that although an inventor, upon public use or publication of an invention, has a one-year grace period to file a U.S. application before losing the patent rights, this grace period does not apply to applications in foreign countries. Hence, to avoid losing the option of foreign patents, new developments should remain confidential until a U.S. application is filed.

The Evolution of Software Patent Law

For a long time the patentability of pure software and software/hardware inventions seemed an open legal question. Only recently has U.S. patent law become relatively clear on this issue. Traditional legal principles state that mental processes, mathematical algorithms, and laws of nature are not patentable, and these principles seemed to prevent the patenting of pure software inventions. In addition, before 1983 the federal courts ruled invalid about 75% of all attempts to enforce patents.

The Old Software Patent Cases

The old legal confusion over software patents is apparent in *Gottschalk v. Benson* (cited herein), a 1972 U.S. Supreme Court case that took the view that computer software is little more than a complicated calculator and ruled that a computer process essentially only manipulates numbers. As such, the computer process merely preempted a mathematical algorithm and therefore was not patentable. However, the Court did not rule on whether all computer programs were unpatentable.

In *Parker v. Flook* (cited herein), a 1978 decision by the U.S. Supreme Court, a pure mathematical formula was denied a patent when no physical activity took place and computers and computerization were not part of the claims. Although it rejected the claims of a pure mathematics patent, the Court did say that new and useful computer programs might be patentable, although it reiterated the old principle that an invention that uses a mathematical algorithm cannot be considered new (and patentable) if the only new part is the new mathematics.

In 1981, the U.S. Supreme Court in *Diamond v. Diehr* (cited herein) moved away from the *Gottschalk* holding in an important respect: it acknowledged that computers and software have a function in industrial processes that is unrelated to simple mathematical calculation. Specifically, this case upheld a patent for an industrial process that incorporated computers and computer software even though one of the steps in the process recited a mathematical algorithm. The old anti-software commandment, "thou shalt not patent mathematics," had been eroded. However, this case dealt with a computer software application that was incorporated within a physical process for curing rubber; it did not comment on pure computer programming or data processing patents that did not physically alter matter. Still, with *Diamond* the lower courts went forward with the new pro-patent direction of the law.

In 1983, the new Federal Circuit Court of Appeals was created and given exclusive jurisdiction over patent appeals cases. This court, created to be pro-patent, began to enforce the majority of patents brought before it and tremendously strengthened the legal assumption that all patents issued by the U.S. Patent Office are valid. Furthermore, legal cases over the last 10 years demonstrate that computer inventions, even pure software inventions, can be patentable and enforceable. These inventions have only to meet the statutory patent requirements of being new, useful, nonobvious, and without prior use or publication.

The New Pro-Software Patent Cases

In 1982, the Court of Customs and Patent Appeals (predecessor to the Federal Circuit) ruled in *In re Pardo* (cited herein) that pure programming technique was patentable, in this case a computer compiler algorithm. The court stated that there is a difference between a *mathematical* algorithm (a human mental process) and a *computer* algorithm (not a human mental process). This followed from a 1973 case, *Application of Knowlton* (cited herein), that dealt with a fundamental patent for the programming techniques of relational data bases and focused on the question of what constituted for patent purposes an adequate description of software.

In 1983, the Federal District Court of Delaware carried this trend further in *Paine Webber v. Merrill Lynch* (cited herein), in which the court supported a pure software patent for the Merrill Lynch CMA account and favorably cited the *Pardo*, *Gottschalk* and *Diamond* cases. In *Paine Webber* the court stated that although business methods (including the CMA account) are not patentable, the patent in question merely covered a computerization of a business method, not the method itself. The court therefore refused a motion for summary judgment, reasoning that the patent may cover unpatentable subject matter. The case was a major step forward for software patentability. The court was not concerned that, practically, no mass market brokerage account, such as the CMA account, could be offered and serviced without a computerized system. It permitted a software patent that, practically speaking, wholly preempted the business method that was not patentable. The court took this step even though in *Gottschalk* the Supreme Court seemed to imply that merely computerizing a mathematical algorithm (in a way that wholly preempted the mathematical algorithm) would not be patentable.

Software Patents Without Source Code

Various cases illustrate the interpretation that disclosure of source code is usually not required in software patents. *In re Ghiron* (cited herein), a 1971 federal case, indicates that flowcharts rather than source code are satisfactory so long as the patent application explains each box and arrow in the flow chart, as well as the hardware platform needed to execute the program. In addition, the application should enable the expert reader to implement the software with only a reasonable degree of routine experimentation. (And because such routine experimentation usually permits normal debugging, debugged source code need not be provided.) Thus, when most software development costs lie in coding and debugging source code, a manufacturer may earn a patent monopoly without worrying about giving away the source code to an infringer through the patent itself.

Knowlton, in 1973, dealt with a software patent that did contain a few lines of key sections of source code, but the judges pointed out that the statute required "concise" patents and that they preferred to deal with flowcharts. *Hirschfeld v. Banner* (cited herein), a 1978 federal case, made it clear that a patent claim cannot recite a program without describing it in some manner, but there was no indication that source code was required.

There is some relatively old case law that suggests that in unusual cases, some source code may be necessary in a patent specification to overcome an attack on "enablement" based on "undue experimentation." See *White Consolidated v. Vega Servo-Control*, 713 F.2d 788 (Fed. Cir. 1983). However, this issue appears to have avoided the reported litigation for the last decade, and is probably not a problem in most cases. Indeed, the better law is that actual programming and de-bugging is not undue experimentation at all, especially when considered in light of the great generality of most claims, and considering that most programming is not required to implement the claims but to satisfy the market demand for additional unclaimed features.

1989-1990 Patent Cases

The case law is constantly elaborating on important questions regarding how best to patent software. In particular, cases are developing to clarify the level of required detail of disclosure and to support the patentability of software.

In general, a patent must describe the software invention in sufficient detail to permit the expert reader to implement the invention in its best version. A 1990 federal circuit case, *Northern Telecom, Inc. v. Data Point Corporation* (cited herein), indicates that the amount of required detail of disclosure depends on the difficulty of the software. Hence, the amount of disclosure is a judgment call on the part of the lawyer drafting the patent. (This requires that the lawyer must be knowledgeable in software and computer patent applications.) If the inventive feature of the patent is adequately disclosed at the flowchart level, however, it is usually not necessary to disclose source code (although if the software owner sees no competitive downside, it may help to include it).

An immediately preceding case, *In re Iwahashi* (cited herein), decided by a federal circuit court in November 1989, also supported the patentability of software. The patent here covered a computer program that calculated a coefficient pursuant to a mathematical formula, then stored the answer in memory. This is the type of "calculator" program that might be most susceptible to attack as merely preempting a mathematical formula. However, the court saw this as no problem, reasoning that a mathematical formula was not preempted but instead a device was patented, and upheld the patent. This judgment was based on patent claims that included the physical element of computer memory and the step of storing the calculation results in that memory. Obviously, such a judgment and its implications raise no obstacle for even the most mathematical of programs; the patent lawyer must simply include in the application this step of storing the answer in memory, which is done by any program anyway and would not impose a practical limitation on the invention. Since this case almost eliminates any subject matter barrier to mathematical software patents, the easier case of the nonmathematical patent remains even more secure.

In *In re Grams* (cited herein), a contemporaneous case by the same court, the bounds of a computerized mathematical algorithm patent are further set. Here the court denied a patent where a mathematical algorithm was merely computerized, with the only physical step being the general collection of data to feed the algorithm; no special collection steps were claimed. Therefore, *Grams* tells us that for the mathematical program, data collection and calculation are not enough, but *Iwahashi* tells us that calculation and recording the answer in memory are enough.

The U.S. patent statutes also provide that new and nonobvious decorative designs may be patented for up to 14 years. These design patents are now being used by many software developers to protect distinctive screen displays and features, such as icons for user-friendly formats.

Copyright and 1990 Copyright Cases

Copyright and patent protection can be simultaneously obtained for a software invention. In most ways, however, patent protection, when available, is superior to copyright protection. Copyright protects only against copying of the protected source code, screen display, and command sequence, but not against reverse engineering, whereas patents protect against reverse engineering regardless of what source code is used, and hence can give much broader coverage. Note, though, that patent rights arise only upon application and grant by the Patent Office, whereas copyright originates as a program is written. It belongs to the writer and lasts for his or her life plus 50 years. Novelty is not needed for copyright. Registration is beneficial but not required.

Recent copyright cases are of special interest for software.

Importance of Contractual Provisions for Software Products

Two 1990 cases indicate the importance of proper contractual provisions for software products when dealing with copyright law.

In a case from December 1990, *Gershwind v. Garrick* (cited herein), out of federal district court in New York City, it was held that, in the absence of contractual provisions to the contrary, a consultant hired to produce certain computer graphic animation for a producer was the owner or joint owner of the resulting computer product. As such, in the absence of provisions in his consulting agreement to the contrary, the owner/consultant could use the product without permission from or payment to his client. A 1989 Supreme Court case, *Community for Creative Non-Violence v. Reid* (cited herein), had already indicated that, to be effective, such contract provisions must be carefully drafted. Ownership of any computer software generated through a consultant, an employee, or a supply agreement should be explicitly specified in writing, even when the employer has provided detailed specifications for the product, as was the case in *Gershwind*.

In a federal Fourth Circuit Court of Appeals case of October 1990, *Lasercomb America, Inc. v. Reynolds*, 911 F.2d 970 (4th Cir. 1990), it was held that a programmer's copyright to his program was invalidated because of prohibited anticompetitive clauses (which constituted copyright misused) in a subsequent license agreement for the software. An analogous danger would lie in a license agreement concerning a software or software/hardware patent and containing the same prohibited clauses. This again emphasizes the need for proper drafting of licenses involving patents or copyrights, because one possible penalty for prohibited provisions is a punitive destruction of the entire property. Such a loss of rights does not run solely to the licensees but also returns all subject rights to the public domain.

Lotus and the Failure of Software Copyright

A software patent protects against reverse engineering by a competitor who develops written source code independently, protection that copyright does not provide. Copyright does little more than protect against copying of the literal source code, as Lotus Development Corp. discovered.

The 1990 federal district court case *Lotus Development Corp. v. Paperback Software International*, Copyright L. Rep. (CCH) ¶ 26,595, 15 USPQ2d 1577, 740 F.Supp. 37 (D.C. Mass.

1990) indicates the prevailing inadequacy of copyright alone to protect software.

In this case, Lotus had no patent for its spreadsheet program. (None of the original spreadsheet companies had such patents, to their great detriment.) Lotus sued Paperback for infringement of copyright only. The sympathetic court cut new ground in extending the software copyright in the direction of a patent. It found that Paperback had copied the structure, seque-nce, and organization of the Lotus menu command system, including the choice of command terms, the structure and order of those terms, the presentation on the screen, and certain long prompts. The court further found that what was copied was a copyrightable, nonliteral element of the computer program and that Paperback had committed nonliteral copyright infringement of Lotus 1-2-3. Although Paperback had not copied the actual Lotus source code, it had reverse-engineered Lotus 1-2-3's function to the user with independently developed source code.

After this case, it now appears that anyone is free to develop and sell spreadsheet programs so long as they use different command terms, structures and presentations on the screen. Hence, Lotus 1-2-3, instead of having a 17-year patent monopoly on the entire spreadsheet market with a basic patent on the concept of the computerized spreadsheet, has only a copyright on its command structure and appearance. Presumably, if Paperback changed the command structure by using synonyms (for example, replacing "delete" with "erase" or "replace"), Paperback would be free to continue selling its spreadsheet program.

Software copyright may look even weaker after the U.S. Supreme Court rules on *Lotus v. Borland* (discussed elsewhere herein). This case, now under consideration by the Court, may lead to a ruling that copyright protects only source code and screen displays, and does not protect command structure, or structure, sequence and organization, or look and feel, of software.

Legal Uncertainty, Defensive Patenting, and Infringement Searches

Software patent law and business practice are developing rapidly and bear watching for developments on the key issues. For example, the U.S. Supreme Court has yet to rule on a *Pardo* type case and clearly state that a computer algorithm is distinct from a mathematical algorithm and therefore patentable. The law also remains unclear on how to create valid liens and security interests in patents. Any of these issues could be decided at any time.

One upshot of enforceable software patents is that the major players in the computer industry are rushing to obtain patents on both past and current work. These patents can then be used to inhibit competitors, as sources of royalty income, or in exchange for other patent rights.

This rush to patent is taking place in a regulatory environment in which the Patent Office is granting, in some form, the majority of applications filed. As a result, in some computer fields a profusion of overlapping and redundant patents has sometimes been incorrectly granted to conflicting parties. In some of these fields, therefore, the presumption of validity for an issued patent may be rebutted, or at least challenged. But even in an area where patents incorrectly overlap, it is still useful to have at least a narrow patent covering your specific product in the marketplace.

This situation emphasizes the need to assess the risk of infringing on a competitor's existing patent by doing a patent infringement search before bringing a product to market. (Such a search should when possible be accompanied by an opinion of counsel to reduce the possibilities of a judgment for triple damages for knowing infringement.)

The bottom line conclusions from these developments are several. Increasingly, a portfolio of software and software/ hardware patent assets will become an important part of the total assets of technology companies, and can be made a profit center through an active licensing program. Further, where substantial sums have been spent to develop software or software with hardware, an additional $10,000 or so to obtain a patent may add excellent marginal returns by reducing competition, increasing revenues from licensing, and providing chips to trade for other patent rights.

Beyond the immediate impact of these patent law changes on a company's profitability, their long-range implications for many industries are monumental. Because hardware (or computing power and mass storage devices) are becoming less and less expensive, companies that primarily manufacture and sell hardware will find declining profit margins and stiffening competition from regions of the world with lower manufacturing costs. Meanwhile, manufacturers of new software applications that take advantage of this increasingly affordable and available hardware will prosper, particularly when they play the software patent card.

A well-drafted patent can be a profitable cash-flowing asset as well as an effective protection for lucrative software and hardware products.

17

Patents for Software and Smart Equipment

Smart equipment can be patented, even when the old dumb version of the equipment is in the public domain. That is, a new patentable invention can arise when an old device is combined with an old microchip, to yield old hardware that is newly programmed. This can be covered by both hardware and software patents. See, for example, U.S. Patent No. 5,018,945, to Baxter International, Inc., May 28, 1991, for an accurate peristaltic intravenous pump with a built-in microprocessor.

This is a rapidly changing area of the law. Watch for new developments. However, all developments are expected to continue to be favorable to software patents.

To address a specific industry, medical software and computer programs can now be patented, and the patents are enforceable. *See Arrhythmia v. Corazonix*, 22 USPQ2d 1033 (Fed. Cir. 1992). This requires that new medical programs and products (1) be reviewed for patenting, in order to avoid copying by others, and (2) be reviewed for infringement of the patents of others, in order to avoid litigation regarding the prior patents of others.

"Single Use Only" labels on patented disposable medical products can now be enforced by patent law. Unauthorized re-use can constitute patent infringement. But beware of anti-trust pitfalls. *See Mallinckrodt v. Medipart*, 976 F.2d 700 (Fed. Cir. 1992).

However, it is legitimate to "invent around" the patents of others, to develop a non-infringing product, which nonetheless competes in the same market with the patented product. *See Kimberly-Clark v. Johnson & Johnson*, 745 F.2d 1437 (Fed. Cir. 1984).

Be careful how the patent claims are drafted. A disposable component of a larger patented device may be copied and/or reused without infringement. This is a matter of patent writing art. Make disposable items a separate independent claim or patent, where possible. *See Surgical Laser Technology v. Surgical Laser Products*, 25 USPQ2d 1806 (D.C. Penn. 1992).

Watch out for accidental loss of corporate ownership of patents and copyrights to employees and consultants. Without proper written contracts, a corporation will not own the technology that it paid for. *See Aymes v. Bonelli*, 980 F.2d 857 (2d Cir. 1992), and again *Aymes v. Bonelli*, 47 F.3d 23, 33 USPQ2d 1768 (2d Cir. 1995).

Software can be both patented and copyrighted. Do both where possible. Copyright alone fails to adequately protect computer software. *See Lotus Development Corp. v. Paperback Software International*, Copyright L. Rep. (CCH) ¶ 26,595, 15 USPQ2d 1577, 740 F.Supp. 37 (D.C. Mass. 1990).

Infringement of software copyrights in the U.S. can be a federal and state crime. (*See* 18 U.S.C. 2319 and Section 31.05(a)(4) Texas Penal Code.) Employees of infringing companies, including officers and directors, can be liable for prison sentences. See *Schalk v. Texas*, 823 S.W. 2d 633, 21 USPQ2d 1838 (Tex. Ct. Crim. App., Oct. 2, 1991).

Willful infringement of patents (for hardware and software) can make an infringer liable for triple damages. (35 U.S.C. 284) Officers and directors of infringer corporations can be personally liable for these triple damages, even if they did not personally infringe, and did not know of the infringement, and even if they were advised by in-house counsel that infringement did not exist. *Data Products v. Reppart*, (U.S. Dist. Kan., 29 November 1990, Lexis 16330) and *3M Corp. v. Johnson and Johnson, Inc.*, 976 F.2d 1558, 24 USPQ2d 1321 (Fed. Cir. 1992).

Independent patent counsel can often act with a corporation to insulate its officers and directors from possible personal liability for payment of triple damages and jail sentences for corporate infringement. These prophylactic steps, when available, should be taken as soon as possible; but they vary with each specific case and should be discussed in confidence with independent counsel. One such step may be a written opinion of non-infringement from outside patent counsel, provided to the corporation before any infringement. This may be partly effective even where the opinion later ultimately proves incorrect. *See Shamrock Tech v. Medical Sterilization*, 25 USPQ2d 1692 (D.C. N.Y. 1992).

There is further good news regarding this personal liability for corporate civil and criminal infringement. By making infringement more dangerous, these developments make patents and copyrights more valuable for their proper owners.

It is easy to lose ownership to patents and copyrights, especially software patents. Use written contracts giving the corporation ownership of all innovations by employees, consultants and partners. (*See Community v. Reid*, 109 S. Ct. 2166 (1989)).

Otherwise, ownership of your technology may end up in the hands of your employees, not your corporation.

Record written assignments to the corporation for all patents and copyrights. Record them in the Patent Office and Copyright Office. (*See* 18 U.S.C. 261.)

Implement a confidentiality program for all your trade secrets, or else you may lose ownership to them.

It is not clear how to create a lien on intellectual property in the U.S. Follow a dual track, using both the UCC (state law) and federal filing. (*See* 18 U.S.C. 261.)

Title searches for intellectual property in the U.S. cannot be reliable because of deficiencies in the statutes. Proceed with great care when reviewing the ownership of patents and copyrights in a merger situation.

There are many proposals before Congress to further change the U.S. patent laws. This is part of a general trend to make patents and copyrights more valuable and enforceable for all. U.S. patent law will become more like patent law in the rest of the world, and foreigners will be treated more equally with U.S. citizens regarding patent applications and infringement. However, these changes in U.S. law must be watched closely, because they will change the proper procedure for protecting intellectual property rights.

The New Rules in Software Patents

It is well known that patents are available for devices. It is less commonly understood that processes may be patented, including computer processes (that is, software programs and algorithms). Additionally, patentable devices may include traditional equipment improved by incorporating new computer controls or micro-processor chips. Indeed, recent legal developments clarify that pure software patents and hardware/software hybrid patents are available and enforceable, and that patents (where obtainable) are superior to copyright protection for software and hardware/software combinations.

Many of these patents claim and patent, at the macro-flow chart level, software algorithms used by specific programs for specific computer applications, sometimes in association with known hardware elements.

Patent applications of this class of technology at the U.S. Patent Office have been growing at about 30% per year for the last four years.

The New Pro-Software Patent Law

In a case in the Court of Customs and Patent Appeals (the predecessor to the Federal Circuit), *In re Pardo*, 684 F.2d 912, 214 USPQ 673 (CCPA 1982), the court ruled that pure programming technique was patentable, in this case an algorithm for a computer compiler. The court stated that there is a difference between a mathematical algorithm (which is a mental process) and a computer algorithm (which is not something that can be performed by the mind). This follows earlier hints from *Application of Knowlton*, 481 F.2d 1357 (CCPA 1973). This case dealt with a fundamental patent for the programming techniques of relational data bases. Although this case focused on the question of what constituted an adequate description of software for patent purposes, it refused to be offended by the concept that pure software was patentable.

Software Patents Without Source Code

Various court cases pursue the interpretation that disclosure of source code is usually not required in software patents. *In re Ghiron*, 442 F.2d 985 (CCPA 1971), indicates that flowcharts rather than source code are satisfactory in software patents. The patent application should explain each box and arrow of the flowchart and the hardware platform required to execute the program. *Ghiron* states that enablement should be provided to the expert reader by the patent with only a reasonable degree of routine experimentation. It is our judgment that normal debugging usually is permitted as routine experimentation, and hence debugged source code need not be provided. Obtaining a software patent without disclosing source code may be like having your cake while eating it too. Where the bulk of software development costs may lay in coding and debugging source code, one may often earn a software patent monopoly without worrying about the possibility of giving an unscrupulous, but underfinanced, infringer his source code through the patent itself.

Knowlton dealt with a software patent that did contain a few lines of key sections of source code. However, the judges reacted rather negatively to that, pointing out that the statute requires "concise" patents and that they prefer to deal with just flowcharts and not source code.

Hirschfeld v. Banner, 462 F. Supp. 135 (D.C. D.C. 1978), made it clear that a patent claim cannot recite a program without describing the program in some manner in the patent. But there was no indication that source code was required in the patent.

There is some relatively old case law that suggests that in unusual cases, some source code may be necessary in a patent specification to overcome an attack on "enablement" based on "undue experimentation." See *White Consolidated v. Vega Servo-*

Control, 713 F.2d 788 (Fed. Cir. 1983). However, this issue appears to have avoided the reported litigation for the last decade, and is probably not a problem in most cases. Indeed, the better law is that actual programming and de-bugging is not undue experimentation at all, especially when considered in light of the great generality of most claims, and considering that most programming is not required to implement the claims but to satisfy the market demand for additional unclaimed features.

The patentability of software was further supported by *In re Iwahashi*, 888 F.2d 1370, 12 USPQ2d 1980 (Fed. Cir. 1989). The patent in this case covered a computer program that calculated a coefficient pursuant to a mathematical formula and stored the answer in memory. This is the type of "calculator" program that might be most susceptible to attack as merely preempting a mathematical formula. However, the court saw this as no problem and upheld the patent. The court reasoned that a mathematical formula was not preempted, and instead a device was patented. This was because the claims included the sole physical element of computer memory and the step of storing the results of the calculation in memory. Obviously, this requirement for a mathematical software patent is no barrier to even the most mathematical program; the sophisticated patent lawyer must simply include in the patent application this step of storing the answer in memory. This surely is done by any program in any case, and would not represent a practical limitation on the invention. Since this case almost eliminates, as a practical matter, any subject matter barrier to mathematical software patents (if very carefully drafted), the easier case of the nonmathematical software patent remains even more secure.

1992 Patent Case

Arrhythmia Research Technology, Inc. v. Corazonix Corp., 22 USPQ2d 1033 (Fed. Cir. 1992), carefully tracks the evolution of software patent law. In *Arrhythmia*, the court reviewed an issued patent for a software algorithm for analyzing electrocardiographic signals to predict heart attacks. The court held that this is patentable subject matter, overturning a lower court decision to the contrary. The court felt that this was in the mainstream of the trend of the cases on point, and the concurring opinion expressly noted the limitation on the anti-software orientation of some earlier case law.

This situation puts additional importance on the need to do a patent infringement search prior to bringing a software product to market, to assess the risk of infringing on an existing patent of a competitor. (Such a patent infringement search should be accompanied when possible by an opinion of counsel regarding non-infringement, to reduce the possibilities of a judgment for triple damages for knowing infringement.)

An important conclusion from these developments is that, increasingly, a portfolio of software and software/hardware patent assets is becoming an important part of the assets of technology companies. In addition, these patent assets can be made a profit center through an active licensing program. Furthermore, where substantial sums of money have been spent to develop software or hardware with software, an additional cost in the order of $10,000.00 to obtain a patent may add excellent marginal returns by reducing competition, by increasing revenues from licensing, and by providing chips to trade for other patent rights. A well drafted patent by a patent attorney knowledgeable in software and computer patent applications can be a profitable cash flowing asset and effective protection for lucrative software and hardware products.

18

Who Really Owns "Your" Patents and Copyrights?

"My best ideas are somebody else's."
-Benjamin Franklin

You may not own "your" proprietary software. Even though your company may have conceptualized the software, written the specifications for it, and paid programmers to write and debug it, you may not own it. The software may be the property of the programmers.

Ownership of software written by employees or consultants can be controlled by the use of properly drafted contracts. However, if contracts are not drawn up and signed, or if contracts are drafted using language that does not reflect recent changes in federal statutes, then the parties that paid for the development of the source code may not own the software.

Or, to put it another way, if you are a programmer or software consulting company that has written a lot of software for clients without written contracts, then you or your consulting company may be the owner of a large library of software assets (which you may have thought belonged to your various clients) and which may be protectable by copyright and patent laws.

Unfortunately, this problem is not academic. In the process of filing patent applications for three new, unrelated clients in a one-month period recently, I discovered that in each case the co-inventors included consultants for whom written contracts had not been drafted. One case involved a new product that was the result of a $10 million R&D effort. In that case, the consultant was reluctant to sign away his newly discovered property without additional payments. All this could have been avoided with a proper consulting contract executed at the beginning of the project.

The legal concepts behind software ownership can be summarized simply enough, but they are not generally understood. Potential legal rights in the area of software can simultaneously include copyrights and patents. A copyright protects the "writings" of "authors" from unauthorized copying. It protects the "words" not the "ideas." A copyright does not have to be applied for (although registering a copyright is a good idea). The copyright arises as the work is created and usually lasts for the life of the author plus 50 years. If an author desires, a copyright can be registered. The original owner of software copyrights, who can be either an individual or a group of people, is the programmer/author. Corporations can obtain eventual ownership to copyrights, however. By contract the programmer/author can transfer and agree to transfer all copyrights by assignment or exclusive license.

A patent protects an inventor from having another person make, use, or sell an invention as his or her won. A patent may be issued to the first inventor, but unlike a copyright, a patent is not automatically granted or created. A patent must be properly applied for and pursued by the inventor. Because of recent legal developments, patents can now be obtained for software programs that are "new, useful, and not obvious." A patent gives the owner a temporary monopoly on the patented software concept or application, which includes the right to prohibit others from creating programs that do basically the same thing but with rewritten, reverse-engineered source code.

The Leading Case

Some older contracts still in use today use language that deems the work of a consultant/programmer to be a "work for hire." Legally, this language attempts to make the copyright to a work, from the time of its creation, the property of the employer of the author, not the property of the author. However, recent changes to the Copyright Act of 1976 (the subject of a 1989 U.S. Supreme Court case, *Community for Creative Nonviolence v. Reid*, cited herein and hereinafter referred to as *Reid*) state that the notion of work for hire applies (with few exceptions) only to employer/employee situations, not to consultant/independent contractor situations. In other words, in cases where a consultant writes software for a client, the consultant, not the client, initially owns the copyright to the software. The consultant will continue to own the software unless he licenses or assigns it in writing to someone else (who may or may not be the client). The work for hire language in a contract does not accomplish this, regardless of the intent of the parties.

In my judgment, *Reid* correctly interpreted the Copyright Act of 1976 as amended. Copyrights are originally owned by the author, but in the case of a work for hire, the employer is considered the author unless ownership is otherwise determined by contract. (*See* 17 U.S.C. 201(a,b).) The statute defines *"work for hire"* as a "work prepared by an employee" and no contract may alter that definition. (*See* 17 U.S.C. 101.) The statute, however, does not define *employee*. In *Reid*, the Court defined *employee*, setting out a 12-factor test. Many (but perhaps not all) consultants would *not* be considered employees, and hence would own their work under this test, unless their contracts (using the proper language) said otherwise.

The Definition of an Employee

The following criteria, derived from *Community for Creative Nonviolence v. Reid*, can used to determine whether a worker qualifies as an employee. No one factor or combination of factors is necessarily determinative.

1. The skill required for the job.
2. The source of the tools used.
3. The location of the work in question.
4. The duration of the relationship between the parties on the project.
5. Whether the paying party had the right to assign work to the paid party.
6. Whether and how much discretion the paid party had over when and how long to work.
7. When and how the price was paid for the work.
8. Whether the paid party could hire and pay assistants.
9. Whether the work was part of the regular work of the paying party.
10. Whether the paying party was in business.
11. Whether employee benefits were provided (and what they were).
12. How the paid party was treated in terms of taxes.

New Software Cases

Reid clarified the copyright statute, but the case actually dealt with the copyright for a statue, not software. Recently, two federal cases have applied the *Reid* test to software, one for the case of a consultant, and one for an employee.

In my view, in *MacLean Associates Inc. v. Mercer-Meidinger-Hansen Inc.*, 21 USPQ2d 1345 (3rd Cir. 1991), the Third U.S. Circuit Court of Appeals correctly applied the *Reid* 12-factor test to a consultant who wrote software, deciding that the consultant, not the client, owned the software copyright. The software program was not a work for hire and the consultant was not an employee, the court said, because the consultant controlled the "manner and means" of how he wrote the software. There was no contract that assigned the copyright. The court went on to state that only the copyright owner may license to copy, distribute, or display the work. (*See* 17 U.S.C. 106.) Furthermore, an effective exclusive license from the owner of the copyright can be made only in writing. (*See* 17 U.S.C. 101, 204(a).)

In *Aymes v. Bonelli*, 980 F.2d 857 (2d Cir. 1992), the appeals court correctly applied the *Reid* test, but reversed the lower court on its application of the facts to the case. The lower court had applied the *Reid* test, and found that an employee programmer did not own the program he wrote because he was an employee. The program was therefore a work for hire and owned *ab initio* by the employer. There was no contract that assigned the copyright to the employee. On appeal, it was found that the court used the correct *Reid* test, but the facts of the case were such that the programmer was not an employee but a consultant. In the absence of a written contract, the programmer owned the copyright, and the company only had an implied license to use one copy. (This would preclude sale of the program by the company.)

But *Aymes* and *Bonelli* were not finished with each other so quickly. In *Aymes v. Bonelli*, 47 F.3d 23, 33 US8Q2d 1768, (2d Cir. 1995), it was recently found that the company was protected under Section 117 of the Copyright Act when it modified its copy of the software, and such a modified copy was not an infringing derivative work.

The lesson to the industry from the *Aymes* litigation is that the company could have saved itself all the litigation with a written contract, properly worded, at the beginning of the programming project, that gave the company all rights to the software.

Since the statute defines *work for hire*, a consulting contract that deems software to be a "work for hire" owned by the client is probably not effective, and ownership of the program would remain with the consultant despite a contract that incorrectly used this term. Where such an unenforceable contract clause is used, the consulting client may have only a nonexclusive license to use the software, and the consultant/creator would be free to sell, license, or use the program with other parties. To avoid this problem, the consultant's client should be careful to draft the consulting contract "to assign and agree to assign all copyrights" to the client, without using the work for hire concept.

A similar situation arises in software with patent rights. The original ownership of a software invention rests with the inventor. Depending on the facts in any particular case, the software inventor may be the programmer, the employer, the consultant client, or others. Where the programmer is the inventor, by written instrument the inventor may assign ownership or exclusively license the patent to an employer or consulting client. In the absence of such an assignment or license, the employer or consulting client may have at best a nonexclusive license to use the patent (sometimes called "shop rights"), but the inventor would retain all other rights to make, use, sell, or license the invention to other parties.

The Solution: Prior Contract

Therefore, whenever an employee or consultant develops software for a company, that company should seek in advance a written agreement that, among other things, specifies where ownership of all the rights, including copyrights and patents, will finally rest. This can usually be done by an assignment of, or an agreement to assign, all intellectual property rights.

Note, however, that many software development contracts that handle eventual ownership deal only with the copyrights. This is a major oversight, since it is now clear that software inventions have the possibility of simultaneous parallel protection by both copyright and patents. Therefore, a software development contract that merely assigns copyrights in a program but does not mention patents leaves the consulting client in the strange position of owning the copyrights to the source code but not the patent rights to the software invention. In this case, the client-copyright owner of the source code could not use it without infringing someone else's patent. In a practical situation, this is as useful as buying half a pair of shoes, and leaves the copyright owner hopping around on one foot with little in the way of useful (or exclusive) ownership rights. A similar "ownership surprise" can arise in cases where software is written by an employee rather than a consultant, although there are some distinctions. The work for hire language will suffice to convey copyright in a contract with an employee, and a written assignment will convey patent rights. But the basic rule holds that without a proper written contract, an employee may own an important interest in the software that he or she writes.

Conclusion

This chapter discusses copyright ownership issues for software. Software may also have overlapping patent ownership issues, which should also be dealt with. The details of patent ownership protection may be distinct from copyright protection, and the patent ownership issues apply to all inventions and patents (including mechanical devices, processes, and drugs) and are not limited to software.

The general answer is that whenever proprietary software (or any new technology) is developed, well-written modern contractual provisions should be used to clarify the ultimate ownership of all intellectual property rights. Otherwise, the parties may be surprised regarding the identity of the actual owner of the new proprietary developments. Indeed, the early drafting and signing of appropriate contracts should be part of a larger program to identify, develop, protect, and exploit intangible intellectual property assets of all kinds.

19

Patents for Software Algorithms

"Whoever invents or discovers a new or useful process ... may obtain a patent therefor ... "

35 U.S.C. 101

"In action there is genius."

-Johann Wolfgang von Goethe

The Court of Appeals for the Federal Circuit has issued a decision that strengthens the legal position that pure software can be patented. The decision, *Arrhythmia Research Technology v. Corazonix Corp.*, 22 USPQ2d 1033 (Fed. Cir. 1992), upheld a U.S. patent that was granted for a "method and apparatus of analyzing electrocardiograph signals" to help predict the likelihood of future heart attacks. The method named in the patent's description is an algorithm that translates an electrical signal into a quantifiable assessment of the patent's cardiac function.

The appellate court's decision overturned the decision of a lower court that had invalidated patent protection for the algorithm. The lower court reasoned that software algorithms are not patentable intellectual property under the U.S. patent statute. But in reversing this lower-court finding, the circuit court held that the previous case law that might have invited this exclusion of software algorithms was limited and did not, in fact, restrain software patentability. The Federal Circuit allowed enforcement of Arrhythmia's patent for its software algorithm.

This new ruling is the most emphatic in a continuing trend in the Federal Circuit to establish the patentability of software and can be expected to lead to more patent applications and legal activity to enforce existing software patents. More importantly, since the Federal Circuit is the exclusive appellate court for patent cases, only the U.S. Supreme Court can invalidate this decision. We can anticipate that this ruling would be supported if ever reviewed by the Supreme Court.

We can predict that this decision will spur more companies and software writers to seek both patent and copyright protection (the latter is much more common and easier to accomplish) and increase legal action to enforce those patents. We can also anticipate increased licensing and cross-licensing activity for software patents, and for software/hardware hybrid patents.

20

How To Bring a New Medical Product to Market

Many successful new medical devices that come to market in the United States are not developed by large medical device companies. Instead, they are initially developed by individual doctors who develop the device to satisfy needs in their own practices, or by medical device salesmen who develop the products in response to a need in the marketplace by doctors or their other customers, or by businessmen who see an opportunity to engage in profitable activity by helping people to satisfy a need.

The question after the initial conception of a new product then is how can a product be brought to market. Only if the product is brought to market will the ultimate user/patient benefit and the original developer experience any financial or personal rewards.

Often, in the case of a doctor or surgeon who develops a new medical product or idea, the idea is simply published in a paper in a medical journal or discussed with medical device suppliers. This approach is easy and gets the idea out to other people, but it results in no financial benefit to the doctor/developer and may not be the best way to get the broadest market penetration to the most patients. In fact, taking this approach, the doctor/developer is in effect giving the innovation away either to the public domain or to the private commercial enterprise that is his supplier.

A Five-Part Product Package

A different approach is recommended to the doctor/developer, which is more likely to lead to market penetration by the product and to financial rewards for the developer. This approach basically outlines a ladder of development of the product to a point where it is ready for commercial introduction into the market, while maintaining the intellectual property rights with the developing doctor for his eventual economic benefit. At some point the product package can be taken by the developing doctor either to a medical company for licensing and sales or to a financial source for the financing of an independent firm to make it and sell it. In either case, the farther the doctor can take the product package along its path of development before he goes to a potential licensee or financial source, the better his chances of success will be in the target transaction.

There are five parts to develop for a new medical product package. These five parts are the engineering development, patents, FDA approval, professional publication, and sales. Each part can be taken part way or all the way down its own development track. The farther along the new product is taken on each one of these five tracks, before the developer takes it to a large corporate licensee or to financing sources to develop an independent company, the better the likelihood of success.

1. Engineering Track

In the engineering development of the product, the steps run from conception of the idea, to designing the product, to making prototype samples, to testing prototype samples, to developing the production design, to securing manufacturing capability. The manufacturing capability, at least initially, can be by a supply contract with the doctor's existing supplier or with some other manufacturing subcontractor.

It is important during this development process that the product be maintained in absolute confidence. The product should not be seen or explained to anyone unless it is absolutely necessary, and then only after obtaining a written confidentiality agreement that has the proper legal language to be enforceable. This is absolutely necessary to keep the proprietary rights to the idea from leaking into the public domain or falling into the hands of a private company or individual other than the developer.

2. *Patenting Track*

Another parallel track in the development of the final product package is to obtain a patent. The ladder of development towards a patent goes from confidential trade secret, to a patent pending (which is really not a patent at all, but simply means that a patent application is pending), to an issued U.S. patent. The maximum protection is provided by an issued U.S. patent in the hands of the developer. However, once a patent application is on file in the U.S. Patent Office, the inventor's eventual proprietary rights are documented with a filing priority date. As a practical matter, at this point the developer can, in most circumstances, disclose the product openly without written confidentiality agreements. This might include discussions with potential suppliers or licensees, publication of technical papers, or descriptions at speeches.

In addition to U.S. patents, some consideration may be given to the economics of investment in parallel foreign patent applications. Also, in some circumstances consideration might be given to patent and copyright protection for any necessary software that may be developed for the device. It is not generally appreciated, but software developments may now be subject to overlapping copyright and patent protection. The cost of filing a patent application varies, but legal fees for drafting and filing most U.S. patent applications probably fall between $5,000 and $15,000, depending on the nature and extent of the invention.

3. The Medical Journal Paper

After the patent application has been filed, there is an excellent opportunity to develop market interest in the product if the doctor/developer publishes a paper in a medical journal describing his use of the product in his practice and the success of the same. This can also be presented at professional conferences on the subject. This method of promoting the product and conditioning the market is particularly effective where the doctor/developer is a practitioner in the field.

4. Sales Track

The ladder of development for sales begins with obtaining marketing and distributor interest. This may entail an initial contractual relationship with a large firm interested in marketing the product with some kind of license and royalty agreement, or with an independent manufacturer's sales representative interested in the same. The next step is development of actual sales history. This establishes that a product can be sold at a price and manufactured at a price and provides an empirical basis for projected business plans indicating cash flow, profitability, and profit margins. The next level in the development of sales is obtaining a backlog of purchase orders to be filled.

5. *FDA Approval*

In the case of a medical device to treat medical conditions, FDA approval is also required. FDA approval for sale follows along two possible tracks, one is the 510K 90-day notice procedure, the other is the PMA human clinical trial procedure. The 510K procedure is for devices that offer no new safety issues to the FDA in comparison with previously existing products. This is the case for many, and perhaps the majority, of new medical devices. The cost of the 510K procedure varies with the product, but may cost between $1,500 and $10,000 per product. The PMA procedure requires human clinical trials and can take years and cost hundreds of thousands of dollars. Therefore, it is essential for an independent developer of a medical product to use creativity and persistence to obtain FDA approval through the 510K procedure, rather than the PMA clinical trial procedure.

Stages of a Business

Many individual doctor-developed products, for example surgical equipment, are taken to the stage of a patent, a prototype, use in an individual medical practice, and publication in a professional medical journal paper, to prepare a successful package for license to a device company. In this approach, no FDA approval or major manufacturing source is obtained, and this is left to the licensing corporation to develop with its own resources. To go to a financing source, FDA approval and a manufacturing source (which may only be a supply contract from a manufacturer) improves the chances of successful financing. Also, of course, taking sales to the point of a backlog of purchase orders is very useful in financing an independent company.

In the stages of financing for a company (from the R&D stage, to the initial market test stage, to the purchase order stage) the lowest business risk for the investor is financing the fulfillment of purchase orders. In this circumstance, there is no R&D or market test left to do. The product has been developed and been successful in the market to the extent that written orders exist. What is then to be financed is merely the working capital for the inventory and overhead to obtain and ship the product to satisfy existing orders. The ideal new product package, then, would entail a patented product with FDA approval, sales history, and a backlog of existing purchase orders.

However, the highest level of development in each of the five tracks is not necessarily required for success. For example, successful licensing may be obtained without a patent, but merely with a patent application pending. In the financial arena, companies are sometimes financed in the R&D phase or in the test market phase before the purchase order phase is obtained. However, the farther along the package is taken in each track, the greater the likelihood of success.

Patent Ownership

It is essential that the owner/developer maintain intellectual property rights for patents and copyrights to the product, otherwise he has nothing of worth to license or finance. Without a patent or the imminent likelihood of a patent, which is solely owned by the inventor, it is unlikely that he will obtain any financial reward from his invention. Of course, in the case of doctor/developers that work for larger institutions, in some cases the arrangements are that the institution owns all or part of the intellectual property rights for developments by the doctor. This does not necessarily prevent the development of the product, as long as the institution is capable of commercializing, licensing or developing an entity to manufacture the product. (Unfortunately, this is often not the case with large institutions, particularly research hospitals, but this is more a matter of organizational culture and orientation of legal right.) However, both the doctor/developer and the institution may want to contribute some care to deal with the question of ownership to get it in the intended party's hands.

We have met with some resistance by medical professionals to the idea of patenting a new medical device. There is an understandable feeling amongst some people that doctors should practice medicine, and when they develop things it should be given to the public for wide use rather than inhibited and tied up in patent rights. Actually, in our capitalist system, reality is exactly the opposite. It is actually more likely that a patented product will obtain wider use in the patient marketplace than an unpatented product. A patented product will deliver more benefit to patients in the end than an unpatented product because a patented product gives an economic incentive to device companies to make the product available and to educate the user public about it. Of course, if a doctor/developer has personal concerns with profiting from medical developments, he can contribute his personal profits to whatever charity he finds most appropriate, after he has commercialized the new development.

21

How Foreign Medical Device Companies Can Penetrate The U.S. Market

Many foreign medical device companies are interested in penetrating the U.S. market because of its size and wealth. However, they are often deterred from attempting the project, particularly for small and medium sized companies for which budgetary restraints may be critical. Over the years in helping a variety of foreign medical companies enter the U.S. market, we have developed an approach that, where applicable, can be customized to help a variety of companies successfully enter the United States market.

Particularly of concern in a project to penetrate the U.S. market is the unfamiliar U.S. legal and regulatory system and the potential expanse of doing business in a market that can be expensive. The essence of this approach is to successfully deal with the regulatory environment in a cost-efficient manner in a way to obtain competitive advantages in the marketplace, and to minimize the financial costs prior to the realization of actual sales. Furthermore, the U.S. liability litigation situation is of concern to many foreign parties unfamiliar with that environment. Steps can be taken in advance to insulate the foreign company to a large extent from this liability and to minimize the overall exposure.

FDA

The first part of the package necessary to enter the U.S. market is FDA (Food and Drug Administration) approval. The sale and use of medical products and drugs in the United States is prohibited without first complying with FDA review for safety and effectiveness. FDA approval can be obtained on two tracks, and the goal is to avoid one track and use the other track. The tracks are the 510K 90-day notice track and the PMA clinical trial track. Devices that are similar enough to existing devices in the U.S. marketplace as to present no new safety issues can follow the 510K track. This track to permission to sell in the United States is relatively fast, cheap and easy. It involves notifying the FDA of intent to sell and providing certain information regarding the product and intended advertising literature.

The alternative track for devices that present new questions of safety and efficacy is the PMA track. This involves human clinical trials in controlled conditions. This is to be avoided if at all possible, since it can take years and costs hundreds and thousands of dollars. The difference between the two tracks in business terms of time and expense cannot be overemphasized, and creativity and effort should be expended to use the 510K approach before the PMA approach is settled on. If the 510K approach is not available or proves to be unsuccessful, then the financial attractiveness of entering the U.S. market should be reviewed, particularly for a smaller or medium sized company. We have found, however, that for many medical device manufacturers with track records of successful products in foreign markets, the 510K approach is often available, particularly if the applicability of this track is creatively and persistently presented to the FDA.

A source of manufacturing that is approved by the FDA is also required for the U.S. market. However, an FDA approved source for medical devices need not be located in the United States. Indeed, a foreign medical company's existing manufacturing facilities may become an approved FDA source to produce and sell products in the United States. The regulations in this regard are detailed; however, they basically require safe sterile manufacturing and detailed recordkeeping to permit lot tracking of products, and recall of defective products when necessary. These GMPs (good manufacturing practices) of the FDA in many circumstances can be complied with little operational change by foreign manufacturers, except perhaps for the institution of additional recordkeeping and lot numbering procedures.

Instruction and assistance in FDA compliance can be obtained from consultants specializing in the FDA, independent manufacturers' sales representatives for the medical industry in the United States, some U.S. law firms, and others. Costs to consultants and attorneys in filing 510K applications for a product vary depending on the product and the consultant, but may run between $1,500 and $10,000 per application.

Import

For foreign manufacturing for U.S. market penetration, import to the United States of foreign manufactured goods is required. Again, a U.S. consultant, customs broker, or law firm can assist in this matter. Certain import requirements and tariffs may be minimized or avoided if parts or sub-assemblies of devices are imported for final assembly in the United States, rather than importing completed parts. This would place the imported goods in the parts category rather than in the medical device category for import purposes. Tariffs in this regard are usually not dispositive of the economics of the question to penetrate the U.S. market, and in general tariffs are expected to go down over the long term, where they exist.

Patents

It is of critical business importance to obtain patents where possible for medical devices to be sold in the United States. A U.S. patent for a product allows the owner of the patent to prohibit competitors from copying the product and gives him the legal right to prevent anyone from making, using, selling, offering for sale or importing the product to the U.S. market. This inhibits competition and allows the patent holder to increase the unit sales of the product, to increase the unit price of the product, and maintain higher profit margins. This is because a patented product can be obtained from only one source, and any competing product cannot be exactly the same. Where a product is new and functionally superior over other products addressing the same problem or market niche, this sole-source patent monopoly can have a dramatic impact on sales and profitability. Indeed, in some instances the fact that a product is patented adds market interest and is taken as an indication of superior performance by the marketplace.

These advantages offered by a patent are important because the U.S. medical market has several huge consolidated players with tremendous market strength. In comparison to a small or medium sized foreign companies, some of these large U.S. companies have, effectively speaking, unlimited financial resources for sales and distribution. The U.S. companies also have famous names recognized among U.S. professional and retail customers. After a foreign company spends significant time and money introducing a product into the U.S. market, and the product has gained enough sales to attract the attention of large U.S. competitors, then the product will likely be copied and sold at a lower price by a U.S. company with a greater market presence and name recognition, unless the foreign company has a patent. In other words, without a patent, a product newly introduced by a foreign company can expect to be crushed by U.S. competition as soon as the product reaches some level of market success. Without a patent, your new product will accomplish little more than a free market test for your U.S. competition.

U.S. Litigation Exposure

The U.S. business climate, in comparison to many other countries, is more litigious, and there is a personal liability lawsuit system that is different from what is found in many other countries. This is an item of some novelty and concern for many foreign medical companies entering the United States market. In this regard, it is important to obtain product liability insurance for new products brought into the United States. Furthermore, consultation with an attorney may provide a legal structure that may dramatically limit the foreign company's exposure to product liability problems in the United States and to insulate the parent company from such exposure. Obviously, there is a large number of small U.S. medical companies that successfully deal with this situation in their own market, and there is an increasing number of small foreign medical companies that are successfully navigating these hazards in carrying on business in the United States. This is not a prohibitive problem, but simply another factor of the U.S. environment to plan for and deal with when beginning a U.S. penetration project.

U.S. Sales

With the preparations described above, U.S. sales are next required. They can be obtained perhaps most quickly, and with the least investment by the small foreign company, by association with U.S. entities. This may entail the use of an independent manufacturer's representative, which is, in effect, a freelance sales organization paid a percentage of their sales. An alternative is to form a joint venture with a U.S. firm or a distributor of U.S. medical products, or a joint venture can be formed with a U.S. medical firm that directly distributes its own products and can do the same for the foreign firm.

It is preferable to design operations and order fulfillment in the United States so as to minimize the need for a large U.S. inventory. This might be accomplished by stocking U.S. inventory from a foreign manufacturing facility or by direct shipment of orders from a foreign manufacturing facility to U.S. buyers.

Furthermore, it may be possible to arrange a private label supply agreement with large U.S. buyers or a U.S. medical device firm. This might particularly be an option with large U.S. firms in the case of a superior innovative patented medical product.

U.S. patent law in several ways does not follow the general trends of law in other developed countries in the world; however, in a manner similar to other countries, to obtain patents for a product in the United States the product must be new, useful, and not obvious. However, the legal test for the amount of newness (often called "inventive step" in the rest of the world) can be surprisingly easy to satisfy. In many instances a smaller level of novelty is required for an invention than might be anticipated by those uninitiated to the U.S. patent system.

Also, of course, before introducing a new device into the U.S. market, it is prudent to check the U.S. patent system to verify that the new device may not infringe on the existing patents of others in the U.S. market. If a newly introduced product infringes patents in the U.S. market, it might expose the company to court orders to stop sales in the United States and requirements to pay monetary damages, either or both of which, of course, will destroy any reasonable business plan.

The Package: A U.S. Business

The basic points of this approach, that is, FDA product approval, FDA approval of manufacturing source, U.S. patents, some level of product liability protection, and U.S. sales, provide a package approach to enter the U.S. market with a minimum obligation of financial expense prior to the achievement of actual profitable results in the U.S. This approach, for example, avoids investment in U.S. manufacturing facilities or U.S. clinical trials, before the market is tested and is generating profits. Front-end expense and commitments are reduced and delayed until after realization of sales, and kept in proportion to actual sales and profits.

These points, once obtained, constitute a U.S. business of financial merit. They constitute a package that merits U.S. equity financing for further expansion, and which obtains superior evaluation for a sale to other larger U.S. medical product players. In particular, U.S. sales at a profit for a patented product will constitute a business with a much higher market evaluation and better merger or financing opportunities, than U.S. sales at a profit for a product with no patent protection that can be copied and crushed by large U.S. competitors at their discretion.

Some consultants, manufacturer's representatives, and entrepreneurial U.S. medical companies may provide assistance to develop this package. They may also provide financial assistance to form a customized U.S. entity to purchase and distribute the products in the United States, once FDA approval and U.S. patents are obtained.

Note that it is interesting that obtaining the FDA 510K approval is not contradictory with obtaining a U.S. patent. In fact, medical products often are patented and get FDA approval by the 510K procedure. The 510K procedure simply requires that a product is not new in a way that will introduce a new safety question. A patent requires newness in an engineering sense, but often this patentable newness does not create a product that presents any new safety issues to the FDA.

Although we have discussed medical devices in this chapter, the approaches described for patenting, FDA applications, and business practices can also be applied to pharmaceutical drugs.

22

Patents as a Profit Center for Hospitals

Hospitals in the United States are beginning to view patents and copyrights as a potential new cash-flowing profit center. These patents and copyrights cover innovations developed by doctors and the staff of the hospital, and can be used to fund the traditional activities of the hospital. Even institutions that are not considered as research hospitals can review their internal management and personnel policies and develop a strategic management plan for capturing and exploiting their hidden intellectual property.

Intellectual property includes patents, copyrights, trademarks and trade secrets, and in the case of hospitals extends primarily to medical devices, new medical treatment methods, software and computer systems (including administrative record keeping and billing software systems customized for hospitals), and data bases (including client lists). Note, in particular, that software applications can now be patented.

Many of these intellectual property programs by hospitals need to emphasize acquiring the legal ownership rights to developments that are routinely generated by hospital personnel and are expensed as part of the normal course of doing business. Hence, the marginal cost of acquiring these ownership rights is relatively small and is the only material new expense of developing these patent programs. Therefore, these programs can become profit centers within the hospitals, when the resulting patents are licensed out. That is, if hospital personnel are generating new developments in technology and administrative methods anyway, then the hospital might as well capture the legal rights to them so that they may be commercially exploited to develop additional financial resources for the institution.

A hospital intellectual property management program should include one or both of the following goals:

1. Generate Cash. The most common use by hospitals of their intellectual property is to generate cash by assigning or licensing the rights to the intellectual property to private companies that can exploit the innovations and pay the hospitals royalties in return. Over time, as an intellectual property portfolio is developed and commercialized, this cash stream can become quite significant. (This technique has been used for some time in certain industrial sectors of the economy and has proven successful. For example, Texas Instruments, which is a semi-conductor manufacturing company, has been aggressively pursuing this basic strategy for over a decade. In recent years the actual revenue to Texas Instruments from the passive collection of patent royalties has exceeded its revenues from the manufacturing and sales of its products.)

2. Protect the Hospital's Competitive Edge. Hospitals, even not-for-profit hospitals, are in competition for the provision of medical services and for funding for medical activity. Although hospitals are not generally in the manufacturing business for medical devices, a patent owned by a hospital for a unique medical device developed at the hospital, or for a unique new method of treatment, or even for trade names, can increase the hospital's ability to differentiate itself from competing hospitals. Likewise, new software patents on administrative systems can further differentiate a hospital's perceived provision of services to its clientele, and may reduce overhead costs. A patent development and licensing program that shares the resulting economic benefits with the hospital and the inventors can also given the hospital a special competitive edge in recruiting and keeping the most creative doctors and staff.

A New Approach

Although hospitals are primarily concerned with the provision of medical services, their staff, medial professionals and administrators are constantly developing new ideas for medical devices, methods of treatment, and administrative software systems, all of which may be patented when new. Traditionally, hospitals as an institution have been unconcerned with the intellectual property rights for these developments. Many hospitals traditionally have let their employees keep any patent or ownership rights that they have developed and not sought to exploit potential markets for their software systems. However, these developments, which naturally occur in the everyday activity of a hospital, can be encouraged by a planned development program. These programs can identify and capture these "natural" opportunities for commercial developments for the benefit of both the hospital and employees involved.

Recycling of Research and Development Funds

Many hospitals, particularly research institutions, have research projects. The funding for these efforts is limited, and to the extent that revenues can be realized from the research, these revenues can be used to make the R&D efforts either self-funding or actual profit centers subsidizing the rest of the institution. This "financial recycling" is a common goal for intellectual property programs of hospital institutions. This approach is even possible for government funded research and development projects. (Often times good negotiation at the beginning of a government funded R&D project can permit the research institution to keep all or most of the ownership rights to any patents eventually developed as a result of the project.)

Attracting and Keeping Personnel

Good strategic management of intellectual property for hospitals should include commercialization of the innovations in question, and a policy of distributing some of the financial rewards to the individuals involved. This can become a competitive advantage for recruiting the best medical personnel, since under such a program the personnel can expect the fruits of their research to be brought to market and to share in the benefits.

Developing a Program: Eight Steps

Developing a strategic intellectual property management program can be accomplished in eight basic steps. The following discussion focuses on patents, but analogous steps apply to copyrights, trade secrets, and trademarks.

1. Obtain Disclosure of the Inventions

The best way for a hospital to encourage employees or consultants to disclose their ideas for inventions is to offer a program of cash incentives, typically a one-time payment or a regularly paid percentage of the income resulting from an invention. In some organizations, patent disclosure forms are distributed periodically as a way of soliciting useful ideas regarding inventions.

2. Review the Disclosed Ideas

This is done by a review board that typically includes technical, R&D, marketing and financial personnel. The board may decide to patent an invention immediately, to keep it a trade secret and develop it further, or to disclose it to the public domain without patenting it (an act that preempts patenting by others).

3. Establish a Confidentiality Program

Staff must be trained in confidentiality procedures. Otherwise normal business practices such as routine use of the development, publishing professional papers, giving speeches, advertising, and press releases may allow ownership of new inventions to lapse into the public domain by default before their exclusive ownership can be acquired by registering patents and copyrights. Even worse, title may come to rest in the hands of employees and consultants who may then exercise exclusive rights in competition with the hospital.

A confidentiality program should contain the following items, adapted to a hospital's individual circumstances:

- Signed confidentiality and noncompetition agreements with all employees requiring that confidentiality be maintained during and after employment and stating that employees are restricted from competition with the hospital.

- A written confidentiality policy document signed by all employees indicating their knowledge of and agreement with the policy.

- A review of all publications and speeches by personnel before such presentations are made public.

Use of dated laboratory or research notes by all technical personnel.

Labeling of all trade secret documents as "confidential," the use of paper shredders for trade-secret trash, locks and other security measures to protect secret information, computer and fax passwords, copy-protects, and phone-line scramblers for modem and fax lines.

Exit interviews with all departing employees reminding them of their post-employment confidentiality obligations.

4. Establish a Licensing Program

A licensing program works best when it is regarded as an independent profit center whose staff is compensated with a percentage of the income they obtain through the licensing of patents, rather than when the program is administered by staff on salary.

5. Establish an Enforcement Function

This step attempts to ensure that no one infringes the hospital's patents and it usually requires constant policing and monitoring of the market in order to challenge invalidly issued patents owned by others. Any enforcement function should also include an attempt at negotiation, so that the hospital may suppress infringement without always having to engage in litigation. In tough cases, however, patent owners may find it economic to litigate against infringers both in the United States and in foreign countries.

6. *Ensure That Title to New Developments Comes to Rest with the Hospital*

Contracts should be signed by all employees and hospital consultants making it clear that all rights to developments are the property of, and shall be assigned to, the hospital. Similar agreements must be obtained from all third-party business associates, including, joint-venture partners and suppliers. This may be a matter of considerable sensitivity, especially for existing employees who have in the past kept the patents for their inventions. This hurdle can best be overcome when the hospital can show that it can successfully use its resources to develop and exploit any inventions, and that the employees will enjoy a share in the financial rewards of the development. Perhaps this commitment to mutual benefit by the institution can best be shown by allowing the individual inventor/employee to take all ownership of a patent for his private development if it is not commercialized by the hospital during an initial period, of say, three years.

7. *Determine in Which Foreign Jurisdictions Patent Applications Will Be Filed*

A U.S. patent application yields only a U.S. patent, which gives rights only within the jurisdiction of the United States. If a company is interested in marketing a device internationally, it should decide in which foreign countries counterpart applications should be filed.

8. *Incentive and Authority*

Two themes must be kept in line for this sought of program to work. They are (1) individual financial incentive, and (2) individual authority.

For individuals to participate in the intellectual property development program and make it work, they must see an individual financial incentive for doing so. Developing and commercializing new products and methods is work, and individuals tend not to undertake that work unless they can see a personal financial reward. To do this, some institutions provide a cash bounty for the various stages in the process. They may include: a payment for submission of an invention idea, a payment if a patent is applied for, and a further payment when a patent is obtained. Furthermore, in addition to these cash bounties, the individual inventor/developer may share in a percentage of the financial benefit obtained by the hospital from the development, in effect making the inventor a partner with the hospital. The institution will find that when cash distributions of a percentage of royalties are made to inventors or the staff of the hospital, that the number and quality of invention submissions by the staff will increase.

Another theme needed to make the program work is individual authority. There must be an individual that has both the personal financial incentive and the authority to make business decision about the licensing of intellectual property so that the property may be commercialized. It is a classic criticism of nonprofit research institutions in this country that the good technology that they develop is often trapped inside the institution, because of the institution's inability to make a business decision about exploiting the opportunity. Some institutions have found that the best way to pursue this is to set up an office or a subsidiary to exploit intellectual property developments. Almost in a venture capitalist mode, the employees of such an office are compensated entirely or in part on a commission or performance basis, that is, with a percentage of their receipts of profits obtained for the hospital.

These two concepts of incentive and authority may represent a change in culture for many institutions. However, these are steps that have been successfully undertaken, even by nonprofit research institutions (one good example of which is the Massachusetts Institute of Technology). The same principles and experiences are adaptable to any institution with new technology that may be interested in, or required to develop, new sources of income.

A Note on Helping the Sick

It is useful to make a comment on the role of profits and the development of medical services. It is thought by many practitioners in the field that new advances in medical devices and methods should be provided free to the widest possible audience and not monopolized even for a limited period by patents for which royalties are charged.

This is seen particularly in certain current controversies about patenting medical procedures. For example, the medical press has recently reported on controversies regarding U.S. patent 5,080,111, by Dr. Samuel L. Pallin, for a "method of making a self sealing episcleral incision," and U.S. patent No. 4,127,118, by Latorre, for a "method of effecting and enhancing an erection," and for other medical methods patents. Many medical practitioners feel that medical treatment method patents threaten to inhibit the free flow of new knowledge among doctors, which has been traditional in modern medicine. Indeed, medical method patents are invalid in much of the world outside of the United States. However, the American view is that medical method patents are necessary to stimulate invention, reward creativity, and to recycle the financial resources necessary to develop innovation.

On a larger scale it might be noted that in our capitalist system, the best way to get new methods and products distributed throughout the marketplace so that their benefits can be enjoyed by the most patients, is by providing economic incentive to private commercial enterprises to spread these innovations. The best way to do that is to patent the methods and products, to provide common business people with the incentive to bring the innovations to as many patients as possible.

In our system, then, the most "charitable" thing to do with a new development is not to let it lapse into the public domain and give it away (where it may languish), but to encourage its use by the system by patenting it. The patent holders can always distribute the royalties from the patent in any charitable manner that they wish, if they have individual qualms against profiting from medical advances. But in this patented and commercial mode, more patients will enjoy the benefits of the technical advance.

23

U.S. Patents for Foreign Companies

The good news is that foreign companies can obtain U.S. patents to give them monopoly patent rights in the United States, and enforce these rights against U.S. companies in U.S. courts. The bad news is that U.S. patent law is in several important respects different from the predominant approach of patent laws in other countries, and what may be conventional practice in other countries may lead to a loss of patent rights or other legal problems in the United States. However, with proper planning, these problems can be accommodated in much the same way as U.S. companies adapt to them.

U.S. Law is Different

There are several basic differences in United States patent law compared to foreign countries. U.S. patents filed in the U.S. before June 8, 1995 last for the longer of 17 years from the issue date or 20 years from the application date. If filed after June 8, 1995, they last for 20 years from the application date, regardless of when issued. U.S. patent applications are not published for opposition by third parties. The U.S. application process is entirely ex parte, that is, it is a secret, private process between the applicant and the U.S. Patent Office. This eliminates any possibility for oppositions by third parties to U.S. patent applications and probably results in some patents being incorrectly issued in the United States. The validity of an issued U.S. patent may be challenged, however, U.S. patent statutes provide a presumption of validity for any issued U.S. patent. The burden is on the challenger to prove invalidity of an issued U.S. patent. Furthermore, priority disputes between two patent applications in the United States are determined by a first-to-invent rule, not a first-to-file rule. That is, if two independent inventors both simultaneously file for the same patent, the inventor that can prove that he invented first will win, rather than the applicant that proves that he filed first in the U.S. Patent Office. This leads to the possibility that any issued U.S. patent may be invalidated by someone that can prove that they invented the invention before the patent holder, although they may have filed the patent application at a later date.

Recently, provisional patent applications have been enabled in the U.S.

Anti-Competitive Practices

The United States has various laws regarding anti-trust and anti-competitive practices that can negatively impact on licensing and other promotional activities for patents in the United States. For example, licensing a patent in a way that provides for royalties as a percentage of sales after the expiration of the term of the patent, can constitute patent abuse. That is, the licensor is attempting to gain for themselves something more than is provided by the patent statute, that is, the licensor is trying to extend the benefits of the patent monopoly beyond the statutory 17-year period. This attempt runs afoul of the U.S. anti-monopoly laws. This patent abuse may allow any third parties who become aware of the license to infringe the patent and avoid enforcement of the patent, even though the infringer may not be a party to the offending license. Likewise, tie-in arrangements and certain cross-licensing arrangements may also be deemed to be contrary to antitrust laws.

The U.S. 102(e) Date (35 U.S.C. 102(e))

Note that when filing a patent application under the Patent Cooperation Treaty ("PCT") and indicating the United States as a designated country, it a good idea to file the PCT application in the U.S. Receiving Office and to include with the PCT filing the U.S. Oath or Declaration of the inventor, together with the fee for the U.S. national stage, (or to file the U.S. patent application directly in the United States as soon as possible). This provides for U.S. purposes a Section 102(e) date equivalent to the PCT filing date, rather than a Section 102(e) date equivalent to the date that the U.S. national stage is entered through the PCT. The Section 102(e) date is not the U.S. priority date but instead the date that allows the patent eventually issued from the U.S. application to be used as a prior art reference to invalidate other U.S. patents. Any delay between the PCT application and entering the U.S. national stage may be critical or fatal for these purposes.

The U.S. International Trade Commission

The U.S. Commerce Department administers an agency referred to as the U.S. International Trade Commission ("ITC"). The ITC acts as an administrative court to adjudicate claims of unfair trade practices under U.S. trade law. Section 337 of the Trade and Tariff Act allows claims in the ITC by U.S. companies against foreign and U.S. companies who use "unfair import practices" when importing into the U.S. One such unfair import practice is to import goods that infringe U.S. patents. Consequently, approximately 80 percent of the cases within the ITC actually litigate the question of U.S. patent infringement by foreign importers into the U.S. market. The ITC is usually required by law to issue a ruling within one year of the time a complaint is brought against a foreign party by a U.S. complaining party. Therefore, the U.S. complaining party has all the time it finds necessary to prepare a case against the foreign importer. However, once the U.S. party files a complaint with the ITC, the entire process, including discovery and judgment, will usually be completed in one year. This is a very accelerated period and can work a hardship on the foreign importer. Consequently, when a foreign importer receives notice that it is a respondent in an ITC complaint, it should immediately contact U.S. counsel and respond to the complaint. If not, the foreign respondent may default in the case and suffer a cease and desist order preventing them from further importing into the United States of the goods in question. In the current GATT negotiations, the U.S. is under some pressure to change or eliminate the Section 337 proceedings before the ITC, but it remains to be seen what effect this pressure may have.

Software Patents

Under relatively new U.S. patent law, it is now clearly possible to patent software and software inventions, including algorithms contained in software and other computer applications. Therefore, when a foreign party has unique software that does something new, it is worthwhile to pursue the possibility of a patent application in the United States. For example, it is generally thought that the original inventors of spreadsheet software in the United States failed to obtain a patent for it merely because they did not apply, under the mistaken impression that software could not be patented. This is currently a market involving about $1 billion in sales a year in the United States alone. The market leader is Lotus which currently has about 40 percent of the market. However, Lotus is thought not to be the original inventors of spreadsheet software, and had the original inventors obtained a patent, it may be that they would have still monopolized this market.

Note, however, that computer software may also be protected in the U.S. by overlapping copyright law, in addition to patents. Hence, to transfer title and licenses in software, it is necessary to handle both the patent rights and the copyrights. If a party obtains patent rights but not copyrights to software, or copyrights but not patent rights, then they may be in a position where they are still unable to use the software.

Title to U.S. Patents

Another important feature of U.S. patent law is that title to U.S. patents is vaporous and can easily be lost to the public domain or to individual employees or consultants, by what may be considered normal business practice. Title to patents in the United States arises originally in the inventor and absent written agreement to the contrary, will remain with the inventor rather than in the hands of his employer or consultant client. A written assignment is required to transfer title to patents, and it must be filed in the public records of the U.S. Patent Office to be effective against third parties without notice. Written assignments of future patents and copyrights and written contracts agreeing to assign future patent and copyrights, with employees, consultants, joint venture partners and other parties are usually enforceable. A written non-disclosure agreement should also be used whenever confidential information is disclosed to any party, including during negotiations with prospective business associates. Patented innovations should be kept secret throughout the world and not put on sale until applications are filed in the U.S. and elsewhere. Under U.S. law, in the absence of a written agreement or a contract from the inventor, it is possible for an employer to direct an employee to develop a certain product within the scope of his employment, and to pay the employee, and to receive the product from the employee, yet the employer may have no or only limited rights to the resulting invention. The same situation can arise between a company and a consultant. Even though the company ordered the invention, paid for it and received delivery, the company will not own the invention in the absence of a written agreement or assignment to the contrary. This may be a particular problem in software and other areas where consultants are commonly used, often without written agreements. This is exacerbated by the lack of general knowledge in the software industry of the patentability of software developments.

The Evolution of U.S. Patent Law

Before 1984, it was more difficult to enforce patents in U.S. courts than today. Some forums had become known for their historical preference to not enforce issued patents. However, in 1984, the law was changed in a way that increased the enforceability, and therefore the value of patents. This led to an increasing number of large lawsuits against patent infringers that have recently resulted in substantial judgments against infringers.

Before 1984, all patent cases appealed from federal district court were appealed to the local federal circuit court of appeals. However, after 1984, all patent cases were appealed only to the federal Court of Appeals for the Federal Circuit ("CAFC") in Washington, D.C. Most of the judges for the CAFC were drawn from the discontinued CCPA (the Circuit Court for Patent Appeals) and are thought to be pro-patent. This new court established an uniformity of patent case law and is thought to be pro-patent, enforcing patents more often than past courts and upholding large cash judgments for infringement. Since 1984, case law has also clearly established that genetically engineered life forms, and computer software, are both patentable. This added two new technologies to the list of candidate technologies for patents.

These trends have greatly increased the value and importance of patents in the U.S. in the last decade. This has caused more patents to be applied for and enforced in the U.S. This trend has also changed the organizations in which intellectual property law is practiced in the U.S.

The Evolving Structure of U.S. Law Firms

Unlike many foreign countries, most U.S. patent practitioners are educated and licensed as general attorneys at law, in addition to being qualified to practice before the U.S. Patent and Trademark Office.

U.S. patent attorneys and agents must have an education in science or engineering, and pass the patent law exam given by the U.S. Patent Office. Patent attorneys, but not patent agents, also must have a legal education and pass the general law exam in at least one state. A patent agent has a license to practice before the U.S. Patent and Trademark Office for patent applications and related matters. A patent agent may not represent clients on any other legal matters, or outside the Patent Office, or in any court. A U.S. patent attorney has a patent license like a patent agent, but a patent attorney may also represent a client on any other legal matters, before other government agencies, and in state and federal courts.

Smaller patent specialty firms tend to limit their practices to pursuing patent applications. Larger patent specialty firms are also involved in patent litigation and licensing. Patent specialty firms do not generally do legal projects involving corporate, securities, mergers and acquisitions, banking, trade, import-export, FDA, FCC, FAA, lobbying, or other legal areas of interest to technology companies.

Larger patent specialty firms tend to be mostly oriented to doing patent litigation, which they see as their primary economic opportunity. Hence, these large patent specialty firms may be less sensitive, or not qualified, to pursue (or integrate into their strategy) non-litigation opportunities to assist their clients. However, very large patent specialty firms may have a unique ability to field a large legal staff on a single litigation matter, in those circumstances where it may be useful.

Increasingly, the trend in the U.S. is for large, diversified, full service law firms to develop their own patent practices, with their own patent attorneys, often by merger with existing patent specialty firms, and sometimes by developing departments slowly in-house.

Diversified general practice firms may also have very large litigation groups, but with some percentage of attorneys who historically have litigated civil cases or business issues unrelated to patents.

One upshot of this situation is that foreign companies pursuing U.S. patents may now obtain "one-stop-shopping" for U.S. legal services. That is, they may deal with one law firm to handle and integrate all legal aspects of their penetration of the U.S. market and protection of their market share. This might include, for example, an integrated program of patent application, licensing, lobbying, regulatory affairs, infringement litigation, and corporate finance. Obtaining and litigating U.S. patents may be only one step of this larger plan, and it can be useful to deal with one legal team to handle the whole picture. Furthermore, some U.S. patent practitioners in diversified firms are now in a position to supervise and coordinate an integrated global program of patent development, exploitation, and enforcement.

U.S. Patent Application Tips

There are several practice tips that are useful for foreign companies filing for U.S. patent applications.

A simple direct translation of a foreign application should not be filed as a U.S. patent application without further review. This review should go to substance and beyond the mere format and form of the application and the claims. A valid U.S. patent must have complete enablement, that is, it must describe to the reader everything needed to implement the invention without undue experimentation. Also, unlike practice in Europe, a U.S. patent must have a complete description of the best mode of the invention. That is, the patent must describe the best manner of implementing the invention known to the inventor at the time of his application. It is grounds for invalidation of a U.S. patent if it can be proved that the inventor or applicant kept secret the best knowledge he had available about his invention, at the time of the U.S. application.

U.S. counterpart patent applications should be updated where appropriate to include the latest developed best mode, which may have been developed after the original parent application was filed. Note that the best mode disclosed in the U.S. patent application must be the best mode at the time of the U.S. counterpart application, not as of the priority date of the parent application.

Unlike Japan, the United States is a very broad doctrine of equivalence. Therefore, very broad general patent claims may be effective and enforceable in the United States, where only more specific individual "picture claims" may be enforceable in Japan. Therefore, it is often useful to aggregate several related Japanese patent applications into one broad U.S. application, and to draft new additional broad claims for the U.S. counterpart application.

Also, unlike Europe and Japan, the U.S. Patent Office is thought to execute bad and incomplete patent searches. Therefore, it is within the power of the patent applicant in the United States to make his patent application stronger by doing his own patent search and fully disclosing the results of the same to the Patent Office during the application procedure. And, of course, all parties involved in the patent application have an obligation to disclose to the Patent Office everything that they know regarding the invention that may be of interest. It may be grounds for invalidating a U.S. patent if material information that was known to a party was withheld from the U.S. Patent Office.

Under U.S. patent law a patent application must be made by the inventors, or assignees of the inventors. Furthermore, in the U.S., a second party cannot normally obtain clear title to an invention except by a written assignment from the individual inventors to the new owners, with the assignment being filed in the U.S. Patent Office public records. For example, without such a written filed U.S. assignment, title to an eventual U.S. patent may remain in the individual hands of the employee/inventor of a foreign corporate applicant.

In the U.S., software inventions can be patented. However, the claims must be drafted in an indirect and arcane way, the best form of which is dynamic and the subject of current litigation. U.S. filings of foreign patent applications for software inventions, and for hardware and software hybrid inventions, may have to be re-written before filing to meet U.S. standards and requirements.

Infringement of U.S. intellectual property rights by foreign corporations can lead to personal liability of corporate officers and directors for civil and <u>criminal</u> penalties. One way to avoid this risk is to obtain in advance a written opinion of non-infringement from independent U.S. patent counsel, prior to making, using or selling a new product or service in the U.S. The process, content, and applicability of these opinions are particular to each case, and should be discussed in confidence with U.S. counsel.

U.S. patent, copyright and trademark laws do not provide for a complete title recording regime for U.S. intellectual property. Consequently, it is not possible to do a reliable title search of public records when buying or licensing U.S. intellectual property. Similarly, it is not clear in U.S. law how to create and foreclose liens on intellectual property. Transactions involving ownership of and liens on U.S. intellectual property should be pursued with this in mind, with care taken that available means are used to lessen or insure around the risks in this area.

Proposed Legislation

There are several changes to U.S. patent law that are currently proposed and pending. These should be monitored although it is difficult to predict if any of these changes will become law, or if they do, at what time this will take place. These proposed legislative changes include changing the U.S. patent system to a "first to file" system from the current "first to invent" system. There is also a proposal to publish U.S. patent applications after 18 months, although they are now secret. New proposals would provide for opposition proceedings regarding published patent applications. There is also a proposal to give "prior use" rights to non-applicants regarding inventions, in competition to the patent rights to the eventual patentee. This proposal may be unconstitutional, however, it appears to have certain legislative support. This proposal seems particularly unworkable and unlikely, although it does receive support in some circles and is embodied in a proposed bill before the current Congress. These legislative developments should, of course, be continually monitored.

Enforcement

Foreign holders of U.S. patents can enforce their patents in U.S. courts against infringers in the U.S., and enjoy the same rights and standing as any U.S. party. Through the courts, a patent holder can obtain a permanent injunction against an infringer against all future infringement, and a cash judgment for past damages from infringement. This judgment for damages can be tripled as a punitive measure where there has been knowing infringement. Also, criminal sanctions against certain types of infringers may apply. This, of course, is a two-edged sword. Foreign owners of U.S. patents can enforce their rights, but they should be careful not to become the target of enforcement by others.

24

Foreign Patents for U.S. Companies

As the United States moves more deeply into the global economy, it is increasingly important for U.S. companies with new technologies to develop patent programs that will protect their inventions throughout the international market. In fact, companies that remain ignorant of foreign patent procedures and apply for patents in the traditional manner, that is, from an obsolete, one-nation perspective, can inadvertently destroy their opportunities to obtain foreign patents.

U.S. patents give rights only within the United States. In order to protect its foreign patent rights, a company's patent strategy and business practices must accommodate foreign rules and deadlines, making sophisticated use of complex regional, national, and global patent application and litigation strategies.

To obtain patent protection in a foreign nation or region, companies must ultimately obtain a patent from the appropriate foreign patent office, and the patent must be enforced in that nation's or region's courts. The good news is that by following proper procedures, U.S. companies can achieve these goals.

Critical Differences

Foreign patent regulations and deadlines are much different from those that apply in the United States. For example, when overlapping patent applications are filed by two independent inventors, a rule must be applied to determine which will be granted the patent. The United States uses the "first to invent" rule, that is, the inventor who can prove that he or she was the first to invent a device and diligently pursue its development will obtain the patent rights. The rest of the world, however, uses the "first to file" rule, that is, the first party to file a patent application in a given jurisdiction is awarded the patent.

In the United States, once an invention is placed on sale, a one-year clock is started. Within one year, the inventor must file a patent application in the U.S. Patent Office, or else the invention will lapse into the public domain forever.

In most other countries, however, there is no grace period. Once an invention is placed on sale anywhere in the world it is barred from patentability and is in the public domain. Japan is an exception to this rule. It has no grace period, but an invention is barred from Japanese patentability only if it is placed on sale in Japan before a patent application is filed there.

Issuance of a U.S. patent also creates a bar to an invention's patentability in other countries, since it is a publication of the invention. Hence, companies should take care to file foreign patent applications in a timely manner, at least before issuance of the U.S. patent. On average, U.S. patents are issued (that is, published by the U.S. Patent Office) about 19 months after the patent application is filed. Companies are typically notified that issuance is imminent three to six months before publication.

The Patent Cooperation Treaty

As the above examples demonstrate, companies that want to sell their products overseas must learn the proper procedures for acquiring foreign patents. Prior to 1970, obtaining foreign patents could be an extraordinarily complicated, time-consuming and expensive procedure. Companies had to file applications directly in each individual country prior to an applicable deadline, or lose their patent rights.

Fortunately, the international application procedure was simplified by the signing of the Patent Cooperation Treaty ("PCT") in Washington, D.C., in 1970. The PCT simplifies the process of filing for foreign patents by providing a single international patent application that establishes a uniform filing date in all PCT countries selected by the applicant. Applicants wishing to obtain foreign patents simply file a single form, checking off the countries from which they wish to obtain patents. This act can eventually initiate examination procedures in the various national patent offices.

The PCT is administered by the World Intellectual Property Organization ("WIPO"), which has headquarters in Geneva and a U.S. branch office in Washington, D.C. WIPO does not issue patents. Rather, it serves as a clearinghouse, processing a company's PCT application and distributing it to the selected national patent offices for examination and issuance. Most of the major countries in the world economy have signed the PCT, including the United States, Canada, Mexico, all the countries of Western Europe, most of those in Eastern Europe, Japan, Australia, New Zealand, South Korea, China, Brazil, and Russia. Much of Latin America, Africa, and the Middle East have not signed. India has not signed as of this writing.

The PCT affords an advantage to U.S. companies that wish to obtain patents from other PCT member countries. Because the United States has a one-year grace period, U.S. companies often test market products before applying for domestic patent protection. Normally, of course, this would create a bar to patentability in most other countries. However, if a U.S. company applies for a U.S. patent before test marketing a device, it can then place its invention on the market with impunity, as long as it files a PCT application within one year of the U.S. filing date. This is because the effective PCT filing date (known as the "PCT convention date" or "PCT priority date") reverts to the U.S. filing date.

If the PCT application is filed more than one year after the original U.S. application, however, the effective PCT date will be the actual date of the PCT filing. This can be disastrous if, in the intervening period, the company has put the product on sale, or if another party has filed an application for the same invention.

The European Patent Office

Besides the PCT, there is another simplified method of obtaining patents in Western Europe. The European Patent Convention ("EPC") has been signed by 17 Western European nations as of this writing, creating the European Patent Office ("EPO") and the European Patent. (A European Patent is considered a "national" patent under the PCT system and an application for it can be initiated through a PCT filing, in the same manner as any other national application.) A European Patent has the force of a national patent in any of the EPC countries and is enforced in those countries' courts.

In other words, to obtain patent protection in Western Europe, a company can file separate applications (through the PCT) in the European national patent offices and pursue individual patent examinations, in an attempt to obtain separate patents. Or, it can take a simpler (and less expensive) route and file once through the EPO, pursuing a single patent examination and a single patent that has the effect of all the national patents. Usually, a company will take the EPO route, although it might apply for separate patents if it anticipates any difficulty in examination or oppositions, to avoid putting all its eggs in one basket. Undoubtedly, EPO procedures will evolve as the political and economic structure of Europe changes, so EPO developments should be continuously monitored.

Russia, Taiwan, and China

The Soviet Union no longer exists, and the former Soviet Union patent office in Moscow is receiving and processing patent applications for Russia, which recently signed the PCT. Most of the other new republics from the former Soviet Union have also signed the PCT.

Among the other major manufacturing nations, Taiwan and India have not yet signed the PCT, although they are under pressure from the countries they trade with to do so. Therefore, companies must file patent applications directly in these countries prior to placing inventions on sale, in order to avoid the on-sale bar in those jurisdictions.

China has signed the PCT, but one must speculate on the practical value of a patent for a foreign company in China. Specifically, it is unclear if a foreign patent holder can rely on enforcement of its Chinese patent against a Chinese infringer in a Chinese court. This situation may, however, change over time.

Additional Tips

Understanding the differences between U.S. and foreign patent systems can yield advantageous information for your company. For example, U.S. patent applications are secret until a patent is granted, when the application file becomes publicly available. However, in foreign countries and under the PCT, applications are published about 18 months after the priority date. By monitoring the publication of foreign patent applications, therefore, companies may be able to obtain helpful clues about the pending U.S. applications of their U.S. competitors. This is especially important in cutting-edge technologies, for which a U.S. application may take from two to four years to be issued and published.

There are also ways to achieve timely filings while delaying payment of many required fees. Once a PCT application is filed, a request must normally be made to enter the designated national phases within 20 months of the PCT priority date. This requires companies to timely pay national-phase filing fees and to have their applications timely translated into the various national languages and filed in the respective national patent offices. However, if within 19 months of the PCT priority date, a preliminary examination is requested at the international level, then the deadline for entering the national phase and payment of national-level fees can be delayed until 30 months after the PCT priority date. This enables smaller companies to protect patent rights in foreign countries, while delaying many of the filing costs until they have had time to assess those markets.

Finally, obtaining patents for software can pose problems in certain parts of Europe. In the United States it is clear that patents can be obtained for pure software innovations. The European Patent Office has developed rules similar to those of the United States and will grant patents for new software developments. However, it appears that the courts in the United Kingdom are somewhat more hesitant to uphold patents for software. To obtain U.K. software patents, therefore, it is now probably best to apply to the EPO, designating the United Kingdom as one of the covered countries, rather than filing directly in the United Kingdom.

Three Crucial Steps

In sum, because of the differences between U.S. and foreign patent procedures, it is important for U.S. companies interested in developing overseas markets to take three steps to ensure that they are able to obtain patent protection in foreign countries.

First, because most countries outside the United States award patents to the first party to file a patent application, U.S. companies must keep their inventions confidential until patent applications are filed, and must file U.S. patent applications as soon as possible. Then, within a year, companies must file a corresponding PCT application, which will have the effective filing date of the U.S. application.

Second, because placing a product on sale creates an immediate ban on its patentability in most countries, it is critical that U.S. companies file for at least a U.S. patent (and for a patent from any interesting non-PCT country) before publicly marketing an invention.

Finally, U.S. companies need to be sure to apply for foreign patents before U.S. patents are issued. If a U.S. patent is published before a company files a foreign patent application, most foreign patents will be barred. Of course, the U.S. patent provides no foreign patent protection.

Various proposals presently exist to make U.S. patent law similar to foreign statutes -- for example, to convert the United States to a first-to-file system, and to create a U.S. system that publishes pending patent applications. To date, however, no such laws have been passed.

The Global Patent Plan: Budgets, Markets, and Competitors

There are over 170 countries in the world. Few companies find it worthwhile, or even financially reasonable, to pursue patents in most of them. The practical thing to do is to analyze those foreign markets that are most attractive, and those foreign export oriented competitors that are most dangerous, and then pursue patents in those indicated countries to the extent that budgets allow. The essence of a global patenting strategy in the use of a benefit-cost analysis, on a country by country basis, to pick the proper national targets that will be profitable patent investments.

25

The 1995 GATT Amendments to the U.S. Patent Statute

Another important development in 1995 was the effectiveness of the GATT amendments to the U.S. Patent Statute. These took effect for applications filed after June 8, 1995.

Perhaps the most interesting GATT amendment to the patent statute is the change of the term of a U.S. patent from 17 years from the date of issuance, to 20 years from the date of application, for patents filed after June 8, 1995. (For patents filed before June 8, 1995, the term is now the longer of 17 years from the issue date, or 20 years from the application date.) Foreign priority under Section 119 is not taken into account in this calculation of term.

Since the date of expiration of an eventual patent is now fixed when the application is made, procedural steps that may expedite issuance of patents may now be more popular. In particular, the traditionally obscure procedure for expedited review of patent applications, the so-called Petition to Make Special, may now come into more common use. This is because obtaining earlier issuance of a patent will extend the period of enforceability. Likewise, divisional applications may be more common, as early allowed claims are taken and rejected claims are continued to be argued for.

The shorter patent term may cause "submarine" patent applications, in extreme examples, to surface more quickly, and certainly before they entirely expire. However, the only impact on the submarine patent strategy will be that the submarine can not stay submerged for more than 20 years.

The amendments do not cause publication of pending U.S. patents, or provide for opposition proceedings against pending U.S. applications by third parties. However, these policies are proposed and supported by the Commissioner of Patents. Hopefully, these changes will come about in the United States, but it is not possible to accurately predict when this will happen. Publication of pending applications and permitting third party oppositions (perhaps through expanded re-examinations) would be especially useful in getting more valid and stronger software patents. This is because it is so difficult for the Patent Office or anyone else to do effective prior art searches in the software area. Hence, software patent applicants can look forward to the opportunity of publication and opposition to U.S. software patents, and if and when this happens, U.S. patent practitioners with experience in European and Japanese opposition proceedings will be of particular interest in the United States.

Provisional Applications

The new GATT amendments also allow for provisional applications. The new section 35 U.S.C. 111(b) allows for provisional applications that meet the requirements for an application under Section 112, but lack claims. A complete Section 112 application filed within a year claiming the priority date to the provisional application will be prosecuted with that priority date. This may become a handy procedure in certain circumstances when a publication or on-sale bar may not be avoidable, but a formal complete patent application is not ready for filing.

Foreign Invention Activities

Regarding foreign inventions, a patentee may now provide a date of invention by acknowledged use or other activities in any WTO ("World Trade Organization") member country (over 100 countries have now become WTO member countries). Before this GATT amendment, a date of invention could only be based on U.S. activities, or since 1994 and the NAFTA agreement, activities in Canada and Mexico. This allows inventors applying in the United States to prove first invention by activities outside of the United States.

Regarding infringement, section 271 has expanded the definition of infringement. The old definition provided for that infringement was constituted by making, using or selling the claimed invention in the United States. Now, under the new amendments, offering for sale or importing is included as an infringing activity.

Also, note that other recent amendments provide that if process claims are in a patent, and a product is made by the claimed process but the product is not itself claimed, then using, selling, offering for sale, or importing into the United States an unpatented product made by the patented process will, in itself, constitute infringement. Although this was perhaps written contemplating biogenetic engineering inventions, it may have important implications in some cases involving software claims that assume a process format.

26

The New Federal Trade Secret Crime And Non-Criminal Opinion Letters

For the first time, a federal trade secret crime has been created. (See the *Economic Espionage Act of 1996*, 110 Stat. 3488 (October 11, 1996), amending 18 U.S.C. 1831-1839.) This Act has some unnatural special definitions of important terms, and provides for draconian penalties, and should result in significant changes in practice, and added legal uncertainty, in certain business situations. The Act creates a substantial new deterrent to the theft of trade secrets, not found in the Uniform Trade Secrets Act, particularly for the smaller party aggrieved by a much larger entity.

It is anticipated that this Act will be an important planning point in circumstances such as (1) changes in employment of key personnel (for the former employer, for the new employer, and for the key employees themselves) and (2) for a variety of business transactions such as joint ventures, supply arrangements, and consulting arrangements. To a certain extent, exposure to liability under the Act by the well meaning player may be minimized by a new type of opinion letter from outside counsel, and by good prophylactic practices by the players involved.

In part the statute provides that "whoever, with intent to convert a trade secret... knowingly... without authorization appropriates, ...without authorization copies... destroys, delivers... conveys... receives... or possesses..., knowing the same to have been... appropriated... without authorization, ...or attempts to commit any offense described [herein]... or conspires [to do the same] shall... be fined not more than $5,000,000, or imprisoned not more than 10 years, or both." [Emphasis added.]

Further, the Act has a criminal forfeiture section that provides, in part, that "...in imposing sentence ... [the court] shall order in addition to any sentence imposed that the person forfeit to the United States... any of the person's property used ... in any manner... to commit or facilitate the commission of such violation... [emphasis added]."

This may mean, for example, if a company hires a competitor's employee and is found to later benefit from his inappropriate use of his former employers trade secrets, then the new employer's entire factory that implemented the secret may be forfeited, in a particular case. In doing this, the court may use its discretion and take into consideration "the nature, scope, and proportionality of the use of the property in the offense." The court may use some judgment in mitigating the scope of such forfeiture, but this discretion is inherently unpredictable.

The new federal Act follows in some ways the Uniform Trade Secret Act ("UTSA"), which has been passed in some form in most states. The Act closely follows the UTSA definition of trade secrets, but the Act has no private cause of action or damages to the injured party. The UTSA has no criminal provision, no special definition of owner, and no forfeiture.

Dangerous Definitions

Note that the Act has unnatural special definitions. For example, "trade secret" means "all forms... of information, including... compilations, program[med] devices, ... whether tangible or intangible and whether... in writing, if (a) the owner thereof has taken reasonable measures to keep such information secret, and (b) the information derives independent economic value... from <u>not being generally known or being readily ascertainable</u> through proper means by the public..."[emphasis added]. The most striking part about this definition is that a "trade secret" does not need to be a secret at all. Instead, a trade secret may be no more than "not generally known to or readily ascertainable through proper means by the public." That is, a trade "secret" may in fact be known and ascertainable by the public. In other words, this theft of trade secret crime can send people to jail and fine them millions of dollars for the theft of something that is not even a secret.

Furthermore, the term "owner" is defined as "the person or entity in whom or in which rightful legal or equitable title to or license in the trade secret is reposed." In other words, not only is the trade secret unnaturally defined (to include non-secrets) but the term owner is defined in a way to include licensees and others who have a right to use. The reference to "or equitable title" may mean that the "owner" under the statute may not appear in any title records and may arise only from an unasserted and contestable equitable claim. In fact, "ownership" may not be clear or commonly known, or ascertainable by a search of any legal records, or even resolved until the conclusion of subsequent litigation of disputes as to equitable title.

It may also be anticipated that this new statute will put added teeth and interest in common non-disclosure agreements. Previously, many non-disclosure agreements could be breached with the liability exposure limited to breach of contact and other collateral claims. However, the case of a breach of a non-disclosure agreement can now expose the breaching party to severe criminal penalties.

Hypothetical Cases

Furthermore, in the case of technical information that may be the subject of patents, or patent applications, the question of whether or not a party is actually using the secret may be a complicated technical question. This technical question may be akin to the arcane determination of patent infringement in the same technical arts.

For example, consider the following fact situation. BIGCO possesses certain software. BIGCO licenses the software for a fifty year period to GIANTCO. Pursuant to license BIGCO delivers to GIANTCO version 1.2 of the software. Five years later GIANTCO is using its own version 3.4 of the same software which it has modified with some cooperative efforts by BIGCO. At that point BIGCO claims that GIANTCO is not properly calculating royalties payments to BIGCO under a complicated and ambiguous formula in the license agreement. BIGCO claims breach and demands that GIANTCO cease and desist using the software. GIANTCO responds that it has incorrectly interpreted the license royalty formula and has in fact overpaid BIGCO rather than underpaid it, and demands refunds of the overpayments. Furthermore, GIANTCO concludes that the software it is currently using has evolved so much that GIANTCO no longer uses at all the software licensed under the original license agreement, and that under the terms of the agreement, the subsequently developed software is entirely the property of GIANTCO. However, GIANTCO feels that it would like to continue using, in its branch office in Nome, Alaska, the original release 1.0 under the license. But, GIANTCO feels that BIGCO has repudiated the license and

BIGCO itself breached by not providing continuous support to the software installation in GIANTCO's Nome, Alaska office. BIGCO sues GIANTCO for breach, and GIANTCO counterclaims.

At this point, under the new Act, it is unclear whether a trade secret is involved at all and who is the owner of the same. The resolution of the question involves issues of contract law, and software technology, that are not particularly unusual fact patterns for a contractual dispute. However, one must wonder if GIANTCO is off the hook for crimes under the Act, for not intentionally and without knowledge acting, because it can no readily interpret the Act with full knowledge of the facts.

Another similar ambiguous situation may arise where XYZCO, a software company, hires Mr. Ate Bit Code, the Senior Vice President from a new product development from a competitor, ABCCO. What software techniques does Mr. Code know as part of his general professional expertise and have a right to use on behalf of any employer, and what software techniques does Mr. Code know that are proprietary to his employer? Furthermore, what is Mr. Code exactly doing in his new office and who is capable of ascertaining or reviewing the possible proprietary nature of the same?

(Although we discuss here largely technology trade secrets, trade secrets can be entirely non-technical information, such as customer lists and private terms of trade.)

Prophylactic Opinion Letters and Practices

Under the new statute several prophylactic steps in these similar situations may be taken. For example, the statute requires intent and knowledge in order to accrue criminal liability. If a party that fears exposure presents the facts as they are known to independent outside counsel with the technical expertise to evaluate the circumstances and to offer an opinion of non-liability under the statute, this may provide evidence to establish the lack of intent and knowledge of the crime.

Particularly where the trade secrets involve arcane technology, this is analogous to long standing practice under patent infringement law, regarding opinion letters of non-infringement. In the patent statutes, triple damages may accrue where the infringement is "willful." Case law generally directs that exposure to triple damages may be avoided, even where infringement is found and liability for actual damages and injunctions obtains, if the infringing party attains a prior opinion of non-infringement, from an independent outside patent attorney. The same approach may serve to establish the lack of intent and knowledge in this criminal trade secret statute for similar facts. Unfortunately, the new trade secret statute offers no clear direction in this regard, other than the terminology referring to intent and knowledge.

In the case of change of employment, the new employer may provide written notice and policy statements to new employees and information regarding trade secret statute, and the fact that the company strongly desires avoidance of such acts, and that ultimately it is the obligation of the individual employees to police their own acts. Good practice would suggest that the corporation take some reasonable steps to implement and verify this policy on a regular basis. A paper trail of training of new employees in this regard may be helpful.

For a prior employer of a departing employee, an exit interview may help the company. Notice and information to the employee regarding his ongoing and continuing obligations under the Act, and collateral civil trade secret statutes, and the applicable patent, copyright, and trademark statutes may also be a useful tool.

In particular circumstances where a previous employer has a specific concern, notice letters, modeled something along the lines of an anticipatory cease and desist letter, to the new employer may be useful for the previous employer to encourage new employer to avoid actual violations and to heighten the likelihood or ease approving eventual wilfulness and knowledge. Practice in this may best follow a rule of reason, since overreaching in this area by a previous employer may in extreme cases constitute or be construed as part of a scheme to maliciously harm the legitimate contracts and relationships of the employee and the new employer.

In the last few years some interesting situations have arisen in the United States, prior to passage of the Act, in which high-ranking employees have left employers, and lead to immediate disputes. In these cases, threatened or actual civil litigation moved in parallel with the complaining party encouraging criminal investigations and enforcement activities by U.S. and foreign jurisdictions.

Recent cases of this type have arisen, for example, when the vice president for purchasing left General Motors for Volkswagen, and when the vice president for new software product development left Borland for Symantec.

Now with the passage of the Act, we would anticipate that similar cases may develop in the future in which a complaining party may encourage the U.S. Attorney to initiate criminal investigations and enforcement. To some this represents an inappropriate criminalization of situations that can involve arcane and inherently difficult technical facts. This would seem to provide a destructive level of ambiguity and misdirection to legitimate economic actors.

Indeed, in specific situations, criminal defendants may have a due process defense to the effect that this federal criminal law is unconstitutionally vague considering the inherent difficulty of determining the extent of the boundaries between discrete products and the presence or absence of infringement and ownership. This might be particularly acute in areas of software and new technology industries.

In circumstances where information is easily and commonly copied and transmitted, such as in the Internet and in software industries, a certain level of awareness should be maintained regarding issues of ownership and use of information.

Prospects

It has been suggested by some commentators that this statute may be little used, since the complaining party has no private cause of action and does not stand to obtain direct monetary awards from the criminal action. However, in civil infringement litigation the greatest value to the plaintiff may be in the economic value of an injunction against prospective infringement, rather than the cash award for prior infringement. Under the new criminal statute, the Attorney General may obtain appropriate injunctive relief in a civil action against any violation of the section. (There is no private cause of action for injured parties under the Act.) Furthermore, discovery in a civil infringement suit may be expensive, slow and ultimately unsuccessful for the plaintiff. However, the Attorney General under this criminal act, in the same factual situation, may have superior discovery powers, and pursue a much faster litigation schedule, and in effect, establish in the public record the factual basis for a subsequent civil suit by a private plaintiff under the patent statute, and do it in a manner that is faster, cheaper and more effective then the civil plaintiff may be able to do on his own.

Therefore, this new federal Economic Espionage Act may find considerable use in the future, although it is unclear how practice shall develop at this time. In any case, developments should be watched with interest by industry and appropriate planning steps taken where ascertainable.

27

Ownership, Liens, and Bankruptcy For Intellectual Property

Title defects are pandemic in intellectual property, and it is unclear how to use intellectual property as collateral. This is partly because of ambiguities in the federal patent, copyright and trademark statutes. The statutory ambiguity arises primarily from the incomplete drafting of the title sections of these three federal statutes. Specifically, this creates a lack of any federal statutory determination of whether federal or state law, specifically the UCC, controls questions of liens, security interests, priority, perfection, and foreclosures in intellectual property.

That is, the federal statutes raise the possibility of pre-emption of state UCC law regarding intellectual property (general intangibles under the UCC), but the federal statutes do not explicitly dispose of the question of pre-emption. And if the federal statutes do explicitly pre-empt the state statutes for intellectual property, then the federal statutes do not create an adequate regime for attending to the relevant issues.

Even given this sorry statutory state of affairs, ownership defects in this type of property may also further arise and be perpetuated by use of less than the best practice.

Where Legal Title Arises

With patents, legal title arises with the individual inventor/employee, and not with the corporate employer of the inventor. As a practical matter, safe and secure legal title will remain with these individuals until a written assignment to the contrary is filed with the "assignment" records of the U.S. Patent Office. However, legal and equitable title may transfer privately between parties in an unrecorded insecure manner. Note, in particular, the status of "joint owners" should be avoided as an impractical business arrangement. See 35 U.S.C. 116, and 262.

In copyrights, legal title arises with the author, which is a defined statutory term. See 17 U.S.C. 201(a) and (b). However, in the case of a "work for hire," the author is deemed to be the employer or client of the writer. However, the "work for hire" terminology is tricky and often misused in practice. The term is defined in the statute. See 17 U.S.C. 101 ("work for hire"). Misuse of this terminology may result in a contract regarding ownership of copyright that may not be interpreted by the courts to implement the agreed upon explicit uncontroverted intent of both parties. See *Community of Creative Non-Violence v. Reid*, 490 U.S. 730 (1989).

In trademarks, legal title arises with the applicant for the mark, which is usually the corporate owner. However, beware of the problem of "assignments in gross." That is, a naked assignment of a trademark without a simultaneous assignment of the goodwill of the business in which the mark is used, may result in the mark losing its proprietary nature and entering the public domain.

Practices vary widely regarding the documentation and recording of title for intellectual property in the United States.

Title practices vary even more regarding the creation and foreclosure liens or security interest in these properties.

This whole area has become of increasing interest lately as more of the new business activities of the largest and fastest growing businesses in the United States are based upon intellectual property assets.

No Title Insurance

To date, the situation with title to intellectual property is so problematic that no title insurance industry exists for intellectual property. The statutes currently do not support reliable title searches that could be the basis for title insurance. This is contrary to the circumstances, for example, in real estate, and to a certain extent in personal property. There is no other species of property in this country that suffers from such an inchoate and defective statutory title regime. This damages all technology business, for no good reason.

This matter could be corrected by a relatively simple technical amendment to the federal patent, trademark and copyright statutes. This would facilitate good title and secure liens and foreclosures for intellectual property. This in turn would vastly increase the value of intellectual property in many business transactions.

The Federal Statutes

The federal intellectual property statutes, in dealing with questions of title and liens, are primarily chattel mortgage statutes. The federal statutory title regimes are inchoate, and do not mention basic UCC concepts such as security interests, financial statements, commercially reasonable sales, and so forth. It is unclear on the face of the federal statutes whether they preempt state UCC law regarding these questions in which the UCC is interested.

The key federal patent statute sections on these questions are 35 U.S.C. 261, 262 and 116.

The key federal copyright statute sections are 17 U.S.C. 205(d) and 101 ("transfer").

The key trademark statute section is 15 U.S.C. 1060.

Although the federal statutes are incompetent and do not clearly point out a proper path for creating and foreclosing liens on intellectual property, we can point out better practice, worse practice, and practices guaranteed to fail in certain circumstances.

"Peter Pan" Security Interests

A particular oddity of attempting to interpret the federal statutes on this point is the need to avoid what may be called "Peter Pan" security interests, particularly in patents. That is, the filing of the UCC security interest documents in the federal records may give priority to that UCC security interest and cut off junior interests. However, those same statutes, upon foreclosure of the security interest and transfer of the remainder of the complete fee title, may give the priority date of the fee title as the date of foreclosure, and may not carry back to the date of filing of the security interest and may not cut off intervening interests.

That is, Section 261 of federal statute for patents may provide for a recorded UCC security interest that can never mature from a juvenile security interest into an adult fee title interest, much like Peter Pan never grew up to be an adult. This would be because the federally filed security interest would be unable to cut off junior intervening interests filed prior to the foreclosure sale.

That is, a UCC 9.504(4) reasonable commercial sale upon foreclosure may not under the federal statute deliver fee title that could leapfrog back to the date of the filing of the security interest and cut off intervening interests. On the other hand, if the security interest were filed under on the UCC with the state, it is likely that the UCC is preempted by the federal statute, and does not provide any protection.

The essence of the problem here is that a chattel mortgage statute, like 35 U.S.C. 261, must have explicit language that a junior interest can not cut off a senior interest and a senior interest can not be cut off by a junior interest. However, it is the essence of the UCC (with its modern Twentieth Century statutory invention of the security interest), that the junior transfer of the remainder of the fee title at a foreclosure sale, must cut-off all senior intervening interests back to the date of the lesser estate, the perfected security interest. Therefore, the fundamental operative language of a chattel mortgage statute apparently contradicts the fundamental operative language of the UCC. That is, 35 U.S.C. 261 apparently contradicts UCC 9.504(4). Hence, no case law can easily make the UCC collaborate with the current federal patent chattel mortgage statute.

Open Questions

Fundamental questions that remain unanswered by the current statutory regime include: in questions of security interests and priority and methods of foreclosure, does the federal law or state law govern? That is, do the applicable federal intellectual property statutes preempt the UCC, partially, or totally, or at all?

Another fundamental question is: if federal law applies for filing, what instrument should be filed in the proper federal office? A UCC financing statement? An assignment of all right, title and interest, but for the limited purpose of securing the indicated obligations (a chattel mortgage)?

Another fundamental question that is unanswered is: what is the procedure for foreclosure of a lien or security interest, once one can create it?

Another fundamental question is: what happens after good foreclosure to deficiencies, surpluses, and equitable rights of redemption, if any.

The result of this unfortunate statutory situation is that any use of intellectual property as collateral in the United States is unnecessarily insecure and subject to legal challenge. Opinion letters can offer no comfort in this area.

An informal poll of what is actually being recorded in the U.S. Patent Office in the Assignment Branch, shows a wide variation by the major players in the field.

Best Practice: Patents and Copyrights

Our conclusion is that in the current unfortunate statutory mess, the best practice for patents and copyrights is to file chattel mortgages in the Patent Office and the Copyright Office, respectively, while also complying with the UCC to create a perfected security interest in the collateral.

The federal documents should be present assignments of all right, tile and interest for the limited purpose of securing the indicated obligations. They should not be agreements to assign upon the default of the obligations, since an agreement to assign is probably an executory contract and may be voided in bankruptcy. Instead, the present assignment of the chattel mortgage would in the classic manner currently assign legal but not equitable title.

Upon default one would presume that title would simply remain with the lender. Upon satisfaction of the obligations, the record title could be reconveyed to the borrower.

Unfortunately, the wording of the federal documents, as in the case of many ancient title instruments, should be artfully drafted and form may triumph over substance.

In this simultaneous "belts and suspenders" federal-UCC approach, maybe one approach will work. The UCC and its definition of financing statement is less concerned with formalities than the federal intellectual property regime. The same document that serves as a chattel mortgage in the federal office may also serve to create a security interest in the state, or a state UCC financing statement or UCC-1 form may be used in the state filing.

Best Practice: Trademarks

Regarding trademark, best practice is a little different. Trademarks must be assigned with goodwill. Goodwill is a question of fact and must include the day to day management and policing of the mark and the quality of the goods with which it is used. Most lenders do not wish to actually get into the business of the borrower, even at the level of licensing to the borrower and policing the goods. Consequently, taking a chattel mortgage in a trademark with goodwill is probably not an acceptable business situation, since it would require the lender to be actively involved in the business of the borrower. Instead, probably the thing to do is to file in the UCC regime with the state for the mark and goodwill, and perhaps file a copy of the state security interest document in the federal office, merely to serve notice (with the assumption that the federal filing of the UCC financing statement may be a legal nullity.)

A lien or security interest on a trademark without the goodwill, upon foreclosure, would presumably result in an assignment in gross and the destruction of the proprietary nature of the mark. This is unsatisfactory in almost all circumstances.

Worst Practice

The worst practice in a case of the intellectual property would be to file a UCC form in the states only referencing general intangibles (which may include all intellectual property) under UCC 9.302(3)(a) and 9.302(4). This may be ineffective if there is even a partial preemption of the state law by federal law. Also, a mere reference to general intangibles would probably include trademarks without the goodwill associated with the mark, which would lead to the assignment in gross problem in foreclosure. (However, this approach may work for trade secrets).

The Licensing Subsidiary

A possible alternative in the case of intellectual property, to give more certainty, would be to assign all intellectual property collateral to a wholly owned subsidiary of the borrower. The subsidiary intellectual property owner would then license the intellectual property to the parent. Then a security interest could be taken in all the stock of the subsidiary, which is well regulated by the UCC. This approach is not often done since it requires redesigning the corporate structure of the borrower, and may have some impact on the tax situation of the company in question. However, the structure does avoid the legal uncertainties of taking liens and foreclosing them directly in intellectual property. In most cases, the licensing subsidiary is probably unworkably complex.

The Legislative Solution

The target solution to the current problem is lobbying and legislation to correct the deficient federal patent, copyright, and trademark statutes. This is relatively straightforward in the case of patents and copyrights. In the case of trademarks, it may be that there is something conceptually unassignable about a trademark, and marks may never be good collateral outside of a "package deal" with an entire operating business, even if only a licensing and monitoring business.

The target solution for statutory repair is modeled after the current FAA ("Federal Aviation Administration") statute, which applies to airplanes. There should be by amendments to the federal patent and copyright statute with (1) a single federal site of filing for all title and lien documents, and (2) federal enablement of security interests and foreclosures such that title taken at a foreclosure sale will cut-off intervening interest holders back to the filing date of the security interest. See 49 U.S.C. 1403(a) for the statute providing the scheme for airplanes, which can be a model for patents and copyrights. Also, see *Danning v. Pacific Propeller, Inc.*, 620 F.2d 731 (9th Cir. 1980), *cert. denied*, 449 U.S. 900 (1980).

The American Intellectual Property Law Association ("AIPLA") subcommittee on security interests in intellectual property has an interest in this matter, as does a similar American Bar Association ("ABA") Committee. However, these matters seem to be a low priority issue for Congress and an immediate amendment is not likely. The current bills to amend the patent statute which are likely to pass and become law do not deal with these issues.

Without a statutory fix, we can expect to see further litigation on these issues. Case law can be typified as just beginning to clarify these ambiguous issues. Case law has not reached the stage to adequately dispose of any of these issues. Most of the cases to date on point tend to be only bankruptcy court cases or federal district court cases.

An interesting general discussion that covers many of these points is found with *In re Peregrine Entertainment, Ltd.*, 16 USPQ2d 1017, 116 Bankr. Reporter 194 (D.C. C.D. Cal. 1990). Although this is only a district court case, it was written by the brilliant Ninth Circuit judge, Alex Kozinski, who is always worth reading.

Maximizing Value, Reasonable Sales

In these circumstances it is interesting how one might maximize intellectual property value. In particular, the question arises: what would constitute a UCC 9.504 commercially reasonable sale of intellectual property. Probably the best answer to this question is not a specific number or percentage royalty rate, but instead execution of the procedural steps that would tend to maximize the highest of value for intellectual property.

The law has little comment on this. We could anticipate that such a commercially reasonable sale of intellectual property would have a relatively long "sales" cycle. It would probably be best to pursue a process in which the likely products to be protected by the intellectual property were identified, then the most likely manufacturers of those products were identified, and then the manufacturers are contacted with the information that would tend to encourage a competitive bid situation for the intellectual property. An actual auction proceeding, after some sort of routine public notice, would probably <u>not</u> be useful.

Covenants, Representations and Warranties of Borrower

When a borrower uses patents for collateral to secure a loan or other obligation, certain covenants, representations and warranties that are particular to patents may be used in the loan agreement, and an effective collateral assignment for the patent property should be used and recorded at the U.S. Patent and Trademark Office, to give the lender a certain level of security regarding the maintenance and value of the collateral. Typical language that may be considered and sometimes used in such agreements and assignments, is offered below for consideration. Of course, specific language should be adapted to each individual transaction, and "form" language is particularly unachievable for this type of property.

Covenants

1. Maintenance Fees. Borrower shall timely file all maintenance fees necessary to maintain the Patents. Borrower shall give Lender timely notice of all payment of such fees.

2. Litigation. No infringement litigation involving the Patents shall be undertaken without the prior consent of Lender. The Lender shall control all patent litigation regarding the Patents. All costs of any patent litigation including attorney fees, court costs, settlements and awards shall be the obligation of Borrower, except as otherwise agreed by Lender. Borrower shall immediately notify Lender of any actual or potential litigation regarding the Patents.

3. Re-examination. Any re-examination, re-issue or opposition proceedings in any Patent Office regarding the Patents shall be conducted as determined by Lender.

4. *After Acquired Patents.* Any inventions conceived, Patents applied for, or Patents issued after the date hereof shall be included in the Patent Collateral. Borrower shall notify Lender immediately of any new patent application by Borrower, its date and serial number, and any new patent issued to Borrower. The Borrower shall file any documents necessary in the Patent Office or elsewhere to provide notice of the collateral interest derived herefrom in the Patents. All patent prosecution shall be at the expense of the Borrower. Borrower represents and warrants that all employees and consultants of Borrower have signed written agreements which assign all inventions and patents by such parties to the Borrower.

5. *Assignments.* Borrower shall not license, assign, encumber, offer as collateral, or otherwise dispose of the Patents without the prior written consent of Lender.

6. *Patent Marking.* Borrower shall mark all goods and services covered by the Patents as "patented" or as "patent pending," as appropriate.

7. *Confidentiality and Record Maintenance.* Borrower shall maintain a corporate confidentiality program designed to maintain the trade secret status of the inventions prior to issuance of a Patent for the same. Furthermore, Borrower shall maintain a legal file, invention disclosures, inventor notebooks, and other material relating to the Patents at its headquarters located at ______________________________________, and such records shall be open to inspection by Lender.

8. *Duty to Prosecute.* Borrower shall prosecute to issuance or final rejection after appeal all patent applications for inventions. The cost of prosecution and issuance shall be an exclusive obligation of Borrower. Borrower shall transmit to Lender copies of all official correspondence to and from the Patent

Office. Borrower shall give Lender a power to inspect and copy all confidential Patent Office files pertaining to Patents.

9. Duty to File. Borrower shall file applications for all inventions that are material to the business of Borrower or which may be included within the Collateral to this Agreement.

Representations and Warranties of Borrower

1. Title. Borrower has good, complete, marketable, and unencumbered title to all items of the Patents and the Collateral. There are no security interest outstanding against the Patents or collateral. The schedule of Patents attached hereto as Schedule 1 is complete and accurate and includes all Patents and patent applications in which Borrower has an interest.

2. *Validity and Enforceability.* All the Patents and the Collateral are valid and enforceable, and are not unenforceable due to inequitable conduct or otherwise. Borrower knows of no material prior art, disclosure, public use, or on sale activity, that was not properly disclosed to the U.S. Patent and Trademark Office during prosecution of the Patents. Furthermore, Borrower currently knows of no public use, public disclosure, or on sale activity that could invalidate any Patent or make any Patent unenforceable. The best mode for practicing the invention covered by each Patent is taught in the specification of each Patent.

The Borrower knows of no claim of any third party that any Patent is invalid. The Borrower has not settled any litigation in any way that might invalidate any Patent. Borrower knows of no claim of full or partial ownership of any Patent by any third party.

Borrower represents that all actual inventors of each Patent are duly listed on each Patent.

Borrower warrants that filing the exhibited Collateral Assignment in the records of the U.S. Patent and Trademark Office [and with the Secretary of State of the State of California] creates a valid enforceable collateral interest or security interest in the Patents.

Borrower has no knowledge of any infringement of the Patents by any third party.

Borrower warrants that it has done nothing that could constitute laches or estoppel in an infringement suit involving enforcement of a Patent.

Collateral Assignment

Whereas, _______________ ("Assignor"), a Corporation duly organized under the laws of the State of __________, and having its offices and place of business at ___________________ _____________, owns certain new and useful inventions entitled: "______________________________________"; and

Whereas, _____________ ("Assignee"), is a Corporation duly organized under the laws of the State of ____________, and having its offices and place of business at ___________________ ____________________________: and

Whereas, Assignee and Assignor have entered into a Loan Agreement, dated _______________________, and Assignee is desirous of acquiring, as collateral security for Assignor's performance of Assignor's obligations under the Loan Agreement, the entire right, title and interest in and to the aforesaid inventions and in and to all applications and Letters Patent therefor granted in the United States of America, and in any and all countries foreign thereto:

Now, therefore, to whom it may concern, be it known that, for and in consideration of the sum of One Dollar ($1.00) to Assignor in hand paid by the said Assignee, and other good and valuable consideration, the receipt and sufficiency of which are hereby acknowledged, and solely as collateral security for the performance of Assignor's obligations under the Loan Agreement, said Assignor by these presents does sell, assign and transfer unto said Assignee, its successors, assigns and legal representatives, the full and exclusive right, title and interest for the territory of the United States of America and all countries foreign thereto (including the right to apply for Letters Patent in foreign countries in its own name and to claim any priority rights for such foreign applications to which such applications are entitled under international conventions, treaties, or otherwise), in and to said inventions, and any other inventions now or hereafter conceived which are related to said inventions or the business of Assignor, and in and to all applications for Letters Patent and Letters Patent granted therefor, and all divisions, reissues, continuations and extensions thereof (collectively, the "Collateral").

Assignor hereby authorizes and requests the Commissioner of Patents, and any official whose duty it is to issue patents, to issue all Letters Patent on said inventions or resulting therefrom to said Assignee, or its successors and assigns, as assignee of the entire right, title and interest. Assignor represents and warrants that Assignor has full right to agree to, execute, and deliver this assignment. Assignor agrees that Assignor will without further consideration, but at the expense of Assignee, communicate to said Assignee, or its successors, assigns, or nominees, any and all facts known to Assignor regarding said inventions whenever requested, and that Assignor will execute all divisional, reissue and continuation applications, testify in any legal proceedings, sign all lawful papers, make all rightful oaths, and generally do everything possible for said Assignee, its successors, assigns, and nominees, to obtain and enforce proper patent protection for said inventions in all countries.

In testimony whereof, the Assignor has caused this assignment to be signed by its duly authorized officers and its seal to be attached this______________________day of___________, 19______.

____

Assignor

by:____________

Attest:

____

Secretary

State of ________________________:

: ss

County of ________________________:

On this _______day of __________________, 19____, before me personally appeared ________________ to me known, who being by me duly sworn did depose and say that he is the ________________ of ______________________, the corporation described in and which executed the foregoing instrument; that he knows the corporate seal of said corporation; that the seal affixed to said instrument is such corporate seal; that it was so affixed by order of the Board of Directors of said corporation, and that he signed his name thereto by like order.

____
Notary Public

(SEAL)
My Commission Expires:

____

Intellectual Property Licenses in Bankruptcy

An interesting aspect of title to intellectual property involves certain provisions of the U.S. bankruptcy statute, specifically 11 U.S.C. 365(n). This section deals with the possible rejection by the trustee in bankruptcy of executory contracts, and specifically rejection of unexpired licenses of intellectual property.

Where a company or individual is in bankruptcy, the general rule is that the bankrupt party may reject the unexpired portion of contracts. Where the licensor of intellectual property is in bankruptcy, unexpired licenses for intellectual property would constitute rejectable executory contracts under the general rule, but the general rule is modified by 11 U.S.C. 365(n) for licenses of intellectual property rights.

Under this section, if a trustee for a bankrupt licensor of intellectual property rejects the executory license, then the licensee may elect to either treat the contract as terminated (if the trustee's rejection would constitute a terminating breach under the contract's own terms, or applicable contract law, or otherwise), or retain its rights under the license (except for contract law rights for specific performance) for the duration of the contract and for any period that the contract may be extended on its own terms. If the licensee elects to continue the license for the intellectual property, the licensee must continue to make royalty payments. If the trustee does not reject the license, then the trustee shall continue to perform the license.

The purpose of this section is to insulate a licensee of intellectual property from detrimental reliance on a license from a party that later declares bankruptcy. This is similar to a provision of the bankruptcy statute to protect real estate lessees.

28

How to Make Foreign Patents More Effective in the U.S.

An important case decided by the U.S. Federal Circuit shows how Japanese, German and other foreign patent holders can best act to make counterparts of their home country patents effective in the U.S. In many cases, this may require changes in the foreign patent holders' practice, to avoid common mistakes that unnecessarily may cause them to lose the U.S. patent rights to their inventions.

It is common practice for many foreign patent applicants to draft a parent patent application in their native language, with the content required by their local patent law, and file the application in their local patent office. Then, conventionally, within the one year deadline provided by the Patent Cooperation Treaty ("PCT"), they may file the same application in the United States, after translation to English where necessary, and perhaps also after recasting the foreign claims in a U.S. format.

However, a recent case by the U.S. Federal Circuit Court of Appeals in Washington, D.C., entitled *In re Ziegler*, 992 F.2d 1197, 26 USPQ2d 1600 (Fed. Cir. 1993), makes it clear that this common practice of merely filings translated foreign parent applications in the U.S. Patent and Trademark Office can fail to result in a U.S. patent, or can yield an ineffective, unenforceable U.S. patent. Note that this is a matter of draftsmanship and legal content, and is not a reflection on the nature or quality of the underlying invention.) This case also clearly indicates the correct practice which, if adopted, yields strong enforceable U.S. counterpart patents for foreign parent applications.

In this case, Ziegler and Martin filed their parent German application on 3 August 1954. They filed a translated U.S. counterpart on 8 June 1955, claiming the 3 August 1954 German date as their priority date. Baxter received a U.S. patent for basically the same claims on 19 August 1954. Ziegler argued to antedate the Baxter reference by claiming the German priority date. However, the court refused to give Ziegler the benefit of the German priority date, because the German filing did not meet the U.S. disclosure requirements of 35 U.S.C. 112, including utility, enablement and best mode.

A shocking lesson of this case is that if a foreign filing is to have any impact for priority purposes as a parent application to benefit later U.S. counterparts, the foreign application must meet all the unique disclosure requirements of U.S. patent law. Otherwise, a later U.S. filing by a third party may defeat the U.S. counterpart. Hence, it may be useful to have parent foreign applications reviewed by U.S. counsel before the original foreign filing, to assure that they meet the standards for a U.S. priority date.

The U.S. Patent Office refused to allow Ziegler's application, saying that it was deficient because it did not describe a practical utility for the claimed invention, and did not contain an adequate written description of the claimed subject matter. The Baxter application was allowed and the patent was issued to Baxter. Ziegler appealed the Examiner's decision but lost before the Board of Patent Appeals and lost again, in this case, before the Federal Circuit.

The Federal Circuit ruled that Ziegler was not entitled to the German priority date because the German application failed to disclose a practical utility for the invention and failed to contain a written description of the claimed invention, as required in Section 112 of the U.S. patent statute (even though these were apparently not requirements of the German statute where the priority application was filed). Furthermore, the court ruled that the Ziegler application in the U.S. had the burden of proving its entitlement to the filing date of the German application. Further, the court held that in order to claim the priority of the German application, the German application must contain all the Section 112 disclosure requirements of a U.S. application. The court ruled that the U.S. requirements of utility and enablement must include a description of how to make and use the invention.

Furthermore, the court made it clear that a correct U.S. patent application by another that is filed after the foreign parent priority date, but prior to the U.S. filing of the defective U.S. counterpart, will act as a prior art bar under Section 102(e) to a later corrected U.S. counterpart filing.

U.S. law has several requirements for a successful patent application that are not found in many foreign countries. They include (1) an explicit statement of the utility of the invention and enablement, that describes to the reader how to make and use the invention, (2) a requirement that the claimed invention be adequately described, and (3) a detailed disclosure of the best mode of implementing the invention known to the applicant at the time that the application was filed in the U.S. Many important foreign countries do not require enablement, utility, best mode, or adequate description to the U.S. standards. Therefore, as in the case of *Ziegler*, a simple translation of a foreign parent application filed in the United States may very well fail to yield a good U.S. patent. Even if the application results in a U.S. patent, that patent can be attacked on the grounds of failure of any of these requirements.

There are also other differences between U.S. and foreign patent law. For example, it is said that Japan has a narrow concept of the doctrine of equivalents where the U.S. has a broad concept. Consequently, the U.S. applicants tend to file fewer patents with broader claims, where in Japan applicants tend to file more patents with narrower claims. Hence, Japanese claims translated into English and filed in the United States may be relatively easy to "invent around." It is often useful to take a collection of related Japanese specific patents and redraft them in the form of one broad U.S. application.

For this reason, when foreign companies wish to file U.S. counterparts of their foreign parent applications, they should work with U.S. counsel to draft their applications (both the parent and counterpart application) to meet U.S. disclosure standards. Otherwise, Asian and European applicants may continue to unnecessarily accumulate portfolios of unenforceable patents in the U.S. This is unnecessary since many of the underlying foreign inventions are patentable if properly applied for.

The current effort a global harmonization of the U.S. patent statute may not change many of these differences between U.S. and foreign patent law. The major proposed changes in the U.S. patent statute for harmonization would publish pending patent applications, if proposed, but would not change the Section 112 content requirements for a successful U.S. patent application.

29

Prior Art Combinations and Obviousness

A case from the Federal Circuit undermines the Patent Office's intermittent tendency to reject patent applications on the grounds that the application is obvious in light of a combination of prior art.

See *In re Bell*, 991 F.2d 781, 26 USPQ2d 1529 (Fed. Cir. 1993). In this case, Bell applied for a patent. The application was rejected with the Examiner and later the Board of Patent Appeals, on the grounds that the invention was obvious based on a combination of prior art references. The Federal Circuit overruled the Examiner and the Board of Patent Appeals ruling that the obviousness rejection was incorrect.

The court held that the question of obviousness requires a conclusion of law and is thus reviewed *de novo* by the Federal Circuit. The burden of proving prima facie obviousness if borne by the Patent Office. In carrying this burden, the PTO cannot establish obviousness by "combing the teachings of the prior art to produce the claimed invention, absent some teaching or suggestion [in the prior art] supporting the combination." When an invention is rejected by the Patent Office for obviousness under Section 103 on the grounds that it is obvious in light of the combined teaching of the combined prior art, there must be some teaching or suggestion of the combination in the prior art itself, otherwise the rejection is incorrect. Furthermore, a product claim is not rendered obvious by similarities of its method with methods contained in the prior art.

This is a strong pro-patent pro-applicant case and should be recalled whenever the Patent Office rejects an application because of a combination of the prior art, where the prior art does not teach such a combination.

In re Bell (cited above) is in line with earlier cases on this point. Where the Examiner has rejected claims under 35 U.S.C. 103 as being unpatentable over a combination of prior art, this rejection may be improper. If the Examiner has combined one reference teaching on thing with another reference teaching something else, then such a combination of references may be improper because it is improper to combine references to show obviousness unless something in those references that suggests such a combination. Before references may be combined to show that an invention is obvious, the Examiner must first ask,

> "whether a combination of the teachings of all or any of the references would have suggested (expressly or by implication) the possibility of achieving further improvement by combining such teachings along the line of the invention in suit." *In re Sernaker*, 702 F2d 989, 217 USPQ 1, 5 (Fed. Cir. 1983).

Furthermore, the references cited in combination must themselves teach that such a combination would attack the same problem as the application in examination. That is,

> "[P]rior art references in combination do not make an invention obvious unless something in the prior art references would suggest the advantage to be derived from combining their teachings." Ibid at 6.

Prior art that does not recognize and attack the current target problem, can not be grounds to reject the current application. And, a new application that identifies a new problem can not be obvious in light of any prior art combination that is unaware of the new problem.

> "[T]he discovery of a problem calling for an improvement is often a very essential element in an invention correcting such a problem [citations omitted]. Therefore, since the cited prior art does not appear to have been cognizant of the problem ... it can hardly be said that the references would have suggested [the solution of the present invention]." *Application of Gruskin*, 234 F.2d 493, 498 (CCPA 1956).

Furthermore, even if the references did suggest that they be combined, the Examiner would still have to ask if the present application achieved more than the self-suggested prior art combination. That is, the Examiner would have to ask,

> "whether the claimed invention achieved more than a combination which any or all of the prior art references suggested, expressly or by reasonable implication." *In re Sernaker, supra*, 217 USPQ at 5.

Rejecting software because of a combination of prior art is common. The indicated line of defense by the applicant, then, is particularly useful in many software applications.

30

Internet Service Patents: The 1997 Survey

There is an explosion of patents at the U.S. Patent Office for new Internet services and businesses. These are software patents which are enabled by new developments in U.S. case law and Patent Office regulations that permit and enforce software patents.

These patents are for new Internet services and protect new business models for the Internet. These patents extend to such areas as communication services on the Internet, remote banking, remote shopping, gambling, database search efforts, and new graphical user interfaces ("GUI's").

As Internet commerce and Internet businesses continue their exponential development and growth, we can expect that the new business models that become successful will in the end allocate their market share and market dominance based on the outcome of Internet software patent battles.

Survey of New Internet Service Patents

We recently conducted a survey of U.S. Internet patents, with surprising results. The survey was conducted on June 9, 1997. At the time, 125 issued U.S. patents were indicated as being related to the Internet. Almost all are entirely software patents.

The first U.S. Internet patent issued on September 5, 1989. This is about eight years prior to the survey. However, the number of issued Internet patents doubled since June 25, 1996, or about eleven months before the date of this survey. This clearly indicates an exponential growth in the issuance of Internet patents. Furthermore, we can expect that a much larger number of Internet patent applications are pending before the U.S. Patent Office. Software patent applications today average about three years from application to issuance, so we can expect that most of the growth in Internet applications has yet to come out of the pipeline as Internet patents.

The rate indicated above is doubling every eleven months, so we can expect that by mid 1998, about another 125 Internet patents will issue. Then in the next year to mid 1999, about 250 more Internet patents will issue. Then in the next year before mid 2000, about 500 more Internet patents will issue. This indicates that approximately 1000 Internet patent applications are currently pending in the U.S. Patent Office.

It can be expected that these pending Internet patents will dramatically impact the financial success and market share of today's new Internet related companies. However, the indications are that any new Internet related company, or any company with new Internet related products, should take any steps available to obtain patent protection for their Internet product, or risk being blocked out in the coming patent wars for Internet market share.

This is the general scheme followed in the development of most new U.S. technology based industries, including the automobile, the airplane, the radio, lasers, nuclear magnetic resonance imaging, and a large variety of other industries. That is, initially there is a race to market success and market share. Then follows a shake-out turning on marketing power and finance, for those business models and market niches that have proven to be successful. Then in the final stage of the development of the maturing industry, the remaining players engage in a legal resolution of their respective patent positions to resolve their final market share. It is interesting that the outcome of the final step, resolution of patent issues, is largely pre-determined far in advance by the original patent applications at the onset of business activity by the respective players. That is, by the time the successful business models demonstrate their superiority in the market place, the opportunity for actually filing patent applications for them will have passed. Patent applications must be filed early, or the opportunity is forever lost.

The first issued U.S. patent mentioning the Internet is Patent No. 4,864,559, issued September 5, 1989, and applied for September 27, 1988. This was issued to Digital Equipment Corporation. This was issued within a year of the application, which cannot be expected today for a software patent. The current statistics indicate that the average time from application to issuance of software patents today (including Internet patents), is about three years. One would expect that Internet patents in particular may be taking longer because of an increased growth in this area. It is also interesting that the first Internet patent was issued to a major player, Digital Equipment. The patent is for a method for multicast communication wherein multicast messages are distributed to all nodes in a multicast range. However, this patent references only two prior art patents and five published references, including a paper of October 1983 entitled "Internet Broadcasting."

It is interesting to compare the first Internet patent with the latest U.S. Internet patent in this survey. This is U.S. patent 5,636,292, issued June 3, 1997, and applied for May 8, 1995. First note that this application took over two years to issue, which is probably fast under the current circumstances. It is entitled "Steganography Methods Employing Embedded Calibration Data," and appears to be directed to a computer software technique. The intervening growth of patents in this area are indicated by the fact that this patent refers to 134 specifically named prior art U.S. patent references, 25 of foreign patent prior art references, and 65 other non-patent prior art papers and books. These other references include items from such sources as the Journal of Neuroscience Methods, the Wall Street Journal, and the MIT Media Lab.

(The MIT Media Lab is the professional home of Dr. Nicholas Negroponte, a former professor of the author of this book, and the author of the recent bestseller entitled *Being Digital*. It is indicative of the rapid development of this field that Dr. Negroponte began his teaching career at MIT as a professor of operations research within the civil engineering department. This was, at the time, as direct a career path as any to becoming a founder and director of the MIT Media Lab and a guru of the Internet and PC applications.)

The growth in prior art references alone from the first Internet patent to the most recent Internet patent is indicative of the explosive growth of the technology in the area and the need for new companies to stake out their own patent territory in order to protect their future market niches.

It is also interesting that Patent 5,636,292 is owned by Digimarc Corporation of Portland Oregon. This is a much smaller and newer company than Digital Equipment, the owner of the first

Internet patent. This is typical of the increase in innovative leading edge companies that created and still dominate the Internet business. It is particularly important for start-up companies that may lack deep financial resources and dominant market share, to protect their positions against larger companies with the patent protection available for new Internet products. Otherwise they may get crushed by the "big boys."

A review of the subject matter of the Internet patents today indicates a wide variety of subject areas. There is a variety of specific software communication techniques, particularly involving communication, plus a variety of LAN's, WAN's, and other networks, including traditional telephone and cable networks, and interfaces and gateways between the same.

A variety of specific industries are also dealt with, including electronic commerce, electronic payments, electronic transactions, encryption, public and private encryption keys, the distribution of software with and without fee collection, photographic cameras systems, secure gateways, multi-media transmissions, the sale and distribution of tickets, smart credit cards, near video on demand, transmission of video and audio files through cable systems, multi-casting of digital video servers, web browsers, remote photography, secured data protection and transmission, and e-mail transmission and priority sorting.

Software Patents Generally

Looking at the broader topic of software patents in general, and not just those patents involving the Internet, our recent survey shows some interesting results of major software players active at the Patent Office today. This updated survey was executed on 23 July 1997.

An interesting aspect of this software patent survey is the sharp increase in the role of software patents issued in the last few years to major players in the software industry. This indicates a recent major change of strategy and a growth of activity at the Patent Office for software applications.

Microsoft

For example, Microsoft is indicated as the owner of 313 patents. It is interesting that their first patent was issued May 13, 1986 for "Holder for Storing and Supporting Articles" and did not apply to software. This patent was issued approximately eleven years ago, but half of the Microsoft patent portfolio has been issued since July 23, 1996, that is in the last year. Furthermore, a full 25% of the Microsoft patent portfolio has been issued since February 11, 1997, that is in the last five months before this survey was taken. As in the case with software patents in general, and Internet patents specifically, Microsoft too shows an exponential growth in its software patent activities. We can assume that an even larger growth of Microsoft patents will continue, with even more Microsoft patent applications currently pending in the U.S. Patent Office.

In this light it is interesting to note that Microsoft began life as a software consulting company, and did not enter into the area of proprietary software products until it reached an agreement with IBM to write DOS, on the condition that Microsoft would own the intellectual property rights to DOS in later versions after IBM paid them to write DOS.

Although Microsoft's original patent was not for software, the vast majority of Microsoft's new patent are indeed for software. It is instructive in this light that most of Microsoft's software patents are not for esoteric software programs or techniques, but are addressed instead at software functionality that is perceived and used directly by the individual PC user. This is a good and instructive approach to develop software patents that are of value in the marketplace, by capturing software functionality of recognizable use to the end user. These patents include such topics as the supply of continuous media over ATM public networks, spellchecking, directory services, full text indexes, recognition of handwritten words, user interfaces, graphical user interfaces, video user interfaces, buttons and tool bars, computer mail gateways, concurrent access to databases by multiple users, previewing of computer output, display of program listings, use of path names, non-default dragging operations, use of game ports, dynamic menu reconstruction in graphical user interfaces ("GUI"), data communication among multiple application programs, repairing crosslink clusters in lost directories, and adding on data signals to transmissions of video signals.

We can now expect Microsoft to be a high profile enforcer of its patent portfolio to its immediate financial advantage in the marketplace. The prudent response by actual and potential competitors to Microsoft will be to develop their own patent portfolio, if only for defensive purposes.

EDS

Looking at Electronic Data Systems, EDS, we see similar developments as with Microsoft, but on a less ambitious scale. We note for EDS that they currently have a total of 28 patents, most of which are for software applications. The applications are specific and include fund transfers by automatic teller machines, software for roaming cellular telephones, audio and visual information on demand (video on demand), and remote alarm transmission.

It is interesting to note in this light that EDS still has the reputation of being a software consultant. However, it is clear that they at least have some strategy of moving in the proprietary software direction, that Microsoft had earlier fashioned. This may be an interesting conflict with their current consulting clients, who may seek EDS software consulting services in order to obtain a proprietary competitive advantage for the client only and not for their "consultant." EDS aside, this type of issue is becoming an increasingly interesting item of negotiation between clients and software "consultants," as more software consultants try to follow the Microsoft model and develop their own proprietary software products at the expense of their consulting clients.

EDS obtained its first U.S. patent on January 25, 1977, a full 20 years ago. However, note that EDS has doubled its patent portfolio in the last three years since the end of 1993. This is typical of the acceleration of interest by a software firms in patents.

Netscape

It is interesting in this regard that Netscape at this point does not have any issued U.S. patents. One might expect though that Netscape might have software patent applications pending, in order to secure Netscape's eventual market position versus Microsoft and others. The success of Stac Electronics in *Stac Electronics v. Microsoft Corp.* (cited herein), using this software patent approach, is the subject of a chapter in this book.

Perot Systems

It is interesting that as of the date of the survey Perot Systems is indicated as having no issued U.S. patents.

Oracle Corporation

At the date of the survey, Oracle is indicated as having 14 issued U.S. patents. The last 50% of this patent portfolio was issued after March 1996. All these patents appear to be software oriented patents. All but one of the Oracle patents have been issued since April 1995.

Cisco Systems, Inc.

As of the date of the survey, Cisco is indicated as having 11 issued U.S. patents.

3COM Corporation

As of the date of this survey, 3COM Corporation is indicated as having 63 issued U.S. patents. Half of 3COM's patents have been issued since February 1996. The 3COM portfolio appears to be largely software oriented.

Sun Microsystems, Inc.

As of the date of this survey, Sun Microsystems, Inc. is indicated as having 521 U.S. patents. Approximately half of these patents have been issued since January 1995. The patents cover both hardware and software.

IBM

International Business Machines Corporation received over 1,800 patents in 1996. The survey indicated 2,935 patents issued for IBM as of the date of the survey in the 1996-1997. These patents cover both hardware and software applications.

Cadence Design Systems, Inc.

As of the date of this survey, Cadence Design is indicated as having 30 issued U.S. patents. This portfolio seems to emphasize software but includes hardware features and architecture. About half of the patent portfolio has been issued in the year since April 1996.

Lotus Development Corporation

Lotus Development Corporation is indicated as having 13 issued patents. Half these patents have issued since December 1994. They are software oriented.

Borland International, Inc.

Borland International, Inc. is indicated as having 49 issued patents at the time of the survey. Half of these patents have issued since August 1996. The patent portfolio is software oriented.

Symantec Corporation

Symantec Corporation is indicated as having 6 issued U.S. patents at the time of this survey. The first was issued in November 1993.

Broderbund Software, Inc.

Broderbund Software, Inc. is indicated as having only 2 patents, both issued in 1994.

31

New Developments: 1996 And Early 1997

"The more profound the discovery, the more obvious it appears later."

-Johann Wolfgang von Goethe

In 1996 and the first half of 1997, there were many important new developments in U.S. patent law that effect patents in general and software patents in particular. These developments promoted the trend to strong enforceable patents, including software patents.

PTO Software Patent Guidelines

Early in 1996, the U.S. Patent and Trademark Office issued the final version of its software patent guidelines. See "Examination Guidelines for Computer-Related Inventions," published February 28, 1996 in the *Federal Register*, and effective March 29, 1996. 61 Fed. Reg. 7478 (1996). These guidelines are reprinted in their entirety as Appendix 3 of this book. This final version of the software patent guidelines replaces the proposed software patent guidelines issued June 27, 1995 (and discussed in Chapter 13 of this book).

The final guidelines follow the basic thrust and repeat much of the material of the proposed guidelines. However, there are some interesting changes in emphasis between the final guidelines and the proposed guidelines, which are made without explanation or comment by the Patent Office.

The final version of the software guidelines are like the proposed guidelines in that they promote and clarify the patenting of software inventions. The final guidelines do not dispose of all software patent issues, and the guidelines state that "these guidelines do not constitute substantive rulemaking and hence do not have the force in effect of law." Hence, we can expect continuing development of case law and practice for the remaining issues of software patenting in the U.S. However, the general trend is clear. Software is patentable and is being patented at an increasing rate, and this will continue. Likewise, enforcement of software patents is taking place and this trend will also continue to grow.

The guidelines remove a tradition attack on software patent applications, that is, that software inventions are "methods of doing business" and as such are not patentable. In removing this objection to software patents, the introduction to the final guidelines states, "[Patent] Office personnel have had difficulty in properly treating claims directed to methods of doing business. Claims should not be categorized as methods of doing business. Instead, such claims should be treated like any other process claims, pursuant to these guidelines where relevant." (This is the <u>only</u> comment on this concept in the final guidelines). Conspicuously, the guidelines do <u>not</u> anywhere state that "methods of doing business" are unpatentable.

It appears, then, that the Patent Office continues in rejecting the doctrine that "methods of doing business" are nonstatutory subject matter for patents and hence unpatentable. In other words, the Patent Office directs that "methods of doing business" is not a useful concept in determining patentability, and that method claims should be analyzed for patentability on other bases. This is in line with the general erosion of the "method of doing business" objection to patents. This erosion has been promoted by continuing activities in software patents and other method inventions, as seen in the dissent by Judge Newman in *In re Schrader*, 22 F.3d at 297. (This is a Federal Circuit case by a three-judge panel, which is discussed elsewhere herein.) The dissent urges that the method of doing business exception be abandoned and is "described as error-prone, redundant and obsolete."

Certainly if the PTO had thought that the "method of doing business" attack on patentability were worth anything, then the concept would have been discussed and supported in the software examination guidelines.

Although the final version of the software guidelines do not contain specific examples of computer-readable memory medium claims, the following useful language is found in Part IV.B.1.(a) of the guidelines, "a claimed <u>computer-readable medium</u> encoded with a <u>data structure</u> defines structural and functional interrelationships between the data structure and the medium which permit the data structure's functionality to be realized, and is thus statutory [and is therefore patentable if otherwise satisfactory]" (emphasis added). Furthermore, in the next paragraph, the software guidelines state that "a claimed computer-readable medium encoded with <u>a computer program</u> defines structural and functional interrelationships between the computer program and the medium which permit the computer program's functionality to be realized, and is thus statutory [and is candidate subject matter for a patent]" (emphasis added). These two statements in the guidelines clearly lay out a claim format for data structure and computer program patents. Particularly, these invention types can be claimed as being contained on a computer-readable memory medium, such as a floppy disk, rather than being claimed purely as "a data structure" or "a computer program." This form of claim constitutes an "article of manufacture" and is candidate subject matter for a patent under 35 U.S.C. 101.

The guidelines do not mention financial service patents or accounting methods patents at all, and certainly does not find them to be objectionable types of software patents. Instead, these patent applications will be treated like any other species of software patent and will not be subject to a special "method of doing business" objection. Indeed, the PTO is issuing many patents of this type.

Japan Patent Office Software Patent Guidelines

In February 1997, the Japan Patent Office revised the "Implementing Guidelines for Examining Patent Applications Relating to Computer Software Inventions", for regular applications filed on or after April 1, 1997. The main revisions in these guidelines are that "a computer readable storage medium having a program recorded thereon," and a "computer readable storage medium having data with structure recorded thereon" can be stated in claims as product inventions.

These new guidelines at the JPO are broadly consistent with the final version of the computer software guidelines issued in 1996 by the U.S. Patent and Trademark Office. This may be the fruit of lobbying to harmonize the law on these points by the U.S. PTO and the Japan Patent Office.

It is interesting that at the same time the Japan Patent Office also issued guidelines for a new method for accelerated examination of patent applications in Japan, to be effective after January 1, 1996. This accelerated review is applicable to applications that are being commercially applied or will be commercially applied within two years in Japan, or where the application has counterpart applications pending in countries other than Japan. This may serve to alleviate some of the delay in examination of Japanese patent applications.

These new Japanese software guidelines indicate the growing worldwide importance of software patents, in the United States and throughout the world.

Financial Service Patents: The Pending Case

Shortly after the final version of the software patent guidelines was issued, the Federal District Court decision in *State Street Bank and Trust Company v. Signature Financial Group, Inc.* 927 F. Supp. 502, 38 USPQ2d 1530 (D.C. Mass. 1996) was handed down. This decision is now on appeal before the Federal Circuit. This is the first case about the patentability of financial services since *Paine Webber v. Merrill Lynch* (cited herein), the "CMA" case, discussed herein.

This new District Court case issued a summary judgment that a patent for a type of mutual fund was invalid because it was nonstatutory subject matter. The case involves U.S. Patent No. 5,193,056 entitled, "Data Processing System for Hub and Spoke Financial Services" issued on March 9, 1993. This is a software patent and claim 1 of this patent, the only independent claim, is drawn toward a data processing system for financial services describing a means for computer processing of certain data.

Oddly, this case invalidates the patent, without trial, as a matter of law, on the grounds that it is "a method of doing business" and would "monopolize" the claimed invention. This is done despite the opinion's acknowledgement of the guidelines as a source of guidance on the issue. Clearly, the judge fails to understand the guidelines' rejection of the method of doing business doctrine.

The case is also odd in its invalidation of a patent on the grounds that it monopolizes an invention. This is odd, because it is exactly the point of all patents to monopolize the claimed invention. Indeed, Article I, Section 8, of the U.S. Constitution states, "Congress shall have power ... to promote the Progress of Science and Useful Arts, by securing for limited times to Authors and Inventors the exclusive Right to their respective Writings and Discoveries" (emphasis added).

Being only a District Court case, this case, is not of strong precedential value. It is also quite controversial and many parties expect it to be reversed on appeal. However, the case does indicate the unpredictability of enforcement of financial service patents at this time. However, even in the worst case, if this District Court ruling is upheld, it would apply only against financial service patents and would not reach all software patents.

Furthermore, the independent claim 1 in *State Street* was drafted prior to later developments in the format of software patents (in part manifested by the later issuance of the proposed software patent guidelines and the final version of the software patent guidelines). These later developments would direct additional alternative forms of the patent claims, if the application were filed today, which might circumvent even the broadest reading of this District Court case.

Furthermore, the opinion attacks the patent in question using the Freeman-Walter-Abele test for method patents, despite the fact that the claims are drawn to computer apparatus and not methods. The PTO software guidelines are clear on this point, and state in the introduction "the Freeman-Walter-Abele test may additionally be relied upon in analyzing claims directed solely toward a process ..."

In any case, the development of this case on appeal should be watched with interest.

Design Patents for Computer Icons

The U.S. Patent and Trademark Office issued "Guidelines for Examination of Design Patent Applications for Computer Generated Icons," published April 16, 1996 in the Official Gazette of the U.S. Patent Office, 1185 OG 60 (1996). These guidelines were developed by the Patent Office to assist PTO personnel in determining whether design patent applications for computer generated icons comply with the "article of manufacture" requirement of 35 U.S.C. 171. Basically, the guidelines facilitate design patents for computer icons, if the application is properly drafted, and the icon is otherwise new and nonfunctional. The drawing accompanying the application should show "a computer screen monitor or other display panel or portion thereof," as discussed in 37 C.F.R. 1.152. If the drawing does not depict a computer-generated icon embodied in a computer screen monitor or other display panel, or portion thereof, in either solid or broken lines, then the claimed design should be rejected by the Patent Office for failing to comply with the article of manufacture requirement.

Basically, these guidelines provide predictability and encouragement for this type of design patent, which can be valuable for distinctive icons.

As a practical matter, overlapping trademark and copyright registrations should also be considered for distinctive computer icons.

The Supreme Court and the Doctrine of Equivalents

In 1997, the U.S. Supreme Court reviewed the critical patent law concept of the doctrine of equivalents. This traditional doctrine gives patent holders rights not only to inventions that are actually claimed in patents, but also to all implied inventions that are "equivalent" to the claimed invention. This doctrine greatly increases the value of patents. See *Warner-Jenkinson Company v. Hilton Davis Chemical Company*, 117 S.Ct. 1040, 41 USPQ2d 1865 (1997), reversing and remanding *Hilton Davis Chemical Company v. Warner-Jenkinson Company*, 62 F.3d 1512, 35 USPQ2d 1641 (Fed. Cir. 1995) (en banc).

The biggest news about *Warner-Jenkinson* is what did not happen.

Warner-Jenkinson reviews the doctrine of equivalence in U.S. patent law. Some commentators thought that the doctrine may be subject to major change or total rejection by this Supreme Court case. This did not happen and the basic contours of the doctrine of equivalence remain the same. We can now assume that the doctrine of equivalence will remain unchanged for at least a generation in the United States. The doctrine of equivalence gives patent holders protection not only for the exact invention claimed literally in the patent, but also for a larger circle of "equivalent" inventions that may not be exactly claimed.

The exact test used for equivalence is not defined by the Supreme Court in this case, and the Court indicates that they will defer to the Federal Circuit to establish and maintain the exact test for equivalence. It is left for the Federal Circuit to further indicate whether the test for equivalence in a particular case may be the "triple identity" test (that is to test the identity of function, way and result of the element in question), or the "insubstantial differences" test, or any other test.

The case clearly states that intent plays no role in the application of the doctrine of equivalence. That is, infringement is infringement regardless of the intent of the infringer. There is no "equitable trigger" required to apply the doctrine of equivalence in defining infringement. Because the test for equivalence is objective, intentional copying and minor "designing around" attempts play no necessary role in establishing whether the doctrine of equivalence applies.

On a different point, the case also says that if the patentee fails to explain the reasons for a non-prior art amendment, to a claim during patent claim prosecution, then it will be assumed that the amendment was made to avoid prior art, and file wrapper estoppel will apply to constrain the doctrine of equivalence, as far as that amendment. Therefore, in the patent prosecution phase, wherever claims are changed, it is important for the patentee to state the reasons for the change (particularly if the change has nothing to do with overcoming prior art), in order to avoid possible file wrapper estoppel in the future.

The case also makes it clear that in infringement analysis, the doctrine of equivalence must be applied to all individual elements of the claim and not to the invention as a whole. That is, an infringer must be shown to practice each element of a claim, or that element's equivalent. This ruling controls any over-reaching application of the doctrine of equivalence, and gives a level of protection and security to the "invent around" or "design around" effort.

The case also states that even "trivial" elements cited in a claim are important to find infringement. Therefore, in drafting claims, care should be taken to eliminate nonessential elements of the claim wherever possible, particularly in independent claims. All "trivial" elements are material in looking for infringement, and must be found to be used by an infringer in order to find infringement.

In short, after this Supreme Court review, the doctrine of equivalence is alive and well, but under control. Patents are strong and are to be enforced through the doctrine of equivalence, beyond the literal scope of their claims. But the progress achieved by "inventing around" continues to be supported.

Trovato II

Elsewhere herein we discussed the original Trovato case, *In re Trovato*, 42 F.3d 1376, 33 USPQ2d 1194 (Fed. Cir. 1994) ("Trovato I"). Since our earlier chapter of this book was written, Trovato I was vacated by the Federal Circuit, and the opinion withdrawn. The decision of the Board of Patent Appeals and Interferences regarding Trovato I was also vacated.

In the new second Trovato case, ("Trovato II") *In re Trovato & Dorst*, 60 F.3d 807, 35 USPQ2d 1570 (Fed. Cir. 1995), the Trovato application was remanded to the Patent Office for consideration in light of *In re Alappat*, 33 F.3d 1526, 31 USPQ2d 1545 (Fed. Cir. 1994) and the proposed PTO guidelines for examination of software patent applications. Trovato II represents a further pro-software-patent step by the Federal Circuit, and the overruling of the anti-software-patent three-judge panel decision in Trovato I. This is good news for software patents.

One lesson that we can learn from the Trovato application is not to write software patent applications in completely abstract terms, but to include as much tangible structure as possible when describing an embodiment of a software invention. This will encourage removal of the application from the "realm of abstract ideas," that was of such concern to the panel in Trovato I.

Copyright Protection of Certain Databases

The new U.S. Patent Office Guidelines for Software Patents may also be read in light of a recent Supreme Court case regarding copyright protection for certain databases. See *Feist Publications, Inc. v. Rural Telephone Service Co., Inc.*, 499 U.S. 340, 111 S. Ct. 1282, 113 L. Ed. 2d 358 (U.S. Supreme Court 1991).

This case basically held that the Constitution in Article 1, Section 8, Clause 8 requires some originality as a prerequisite for copyright protection, for databases and compilations or otherwise. Specifically, the Supreme Court threw out the "sweat of the brow" or "industrious collection" test which extended copyright to compilations that represented considerable investment or work by the compiler, but not necessarily any originality. The Court held that investment of time and money into preparing a compilation does not merit copyright protection, and that only originality merits copyright protection, regardless of the time or expense, or lack of time or expense, required in compilation.

In the Feist case, Rural Telephone Service published a typical telephone directory of white pages and yellow pages. The information in the white pages was publicly available, and not trade secrets, but represented a large and expensive compilation. The Rural Telephone white pages did not represent any significant originality. Feist Publications published specialized telephone directories for larger areas than Rural Telephone, but included some of Rural Telephone's area. Feist extracted listings from Rural's white pages telephone directory without Rural's consent.

The Supreme Court held that Rural Telephone had no copyright protection in its white pages information, because Rural's compilation represented no originality, despite the fact that it represented a considerable investment of time and money by Rural. Feist then was free to extract the information from Rural's

Telephone's compilation and use this information in Feist's own competing compilation.

The Supreme Court noted that data compilations could obtain the originality necessary for copyright protection if the author chose the facts in such a way as to constitute some originality. However, such copyright protection would extend only to those components of the work that were original to the author and their presentation, and not to the facts themselves. A compilation is not copyrightable per se, but is copyrightable only if its facts have been "selected, coordinated or arranged in such a way that the resulting work as a whole constitutes an original work of authorship." See 17 U.S.C. 101.

In the specific case in question, the Supreme Court ruled that Rural Telephone had a valid copyright in the directory as a whole because it contained some forward text and some original material in the yellow pages, however, there was nothing original in Rural Telephone's white pages. The raw data was held to be uncopyrightable facts and the way in which Rural selected, coordinated and arranged those facts was not original in any way and typical of all white pages in general. The Court found that the white pages of Rural Telephone lacked the modicum of creativity necessary to transform mere routine selection into copyrightable creative expression.

However, reading the Feist opinion in the context of the PTO patent guidelines for data structure patents, it is clear that even uncopyrightable compilations, if used in a computer in way that provides new and inventive functionality a computer, may be claimed in such a manner as to constitute a larger patentable and protectable apparatus method or article of manufacture. Examples of this may be white pages databases, geographic map information, GPS coordinates, or a vast variety of other data incorporated into a functional software design in an inventive manner.

Digital v. Intel

On May 13, 1997, Digital Equipment Corp. and Cyrix Corp. announced that they sued Intel Corp., accusing Intel of patent infringement of ten Digital patents awarded between 1988 and 1996, allegedly used in the Intel Pentium, Pentium Pro, and Pentium II chips. Cyrix at the same time filed suit against Intel regarding Intel's Pentium line, alleging infringement of two of Cyrix patents. This case has been viewed by some commentators as Digital's attempt to suppress the next generation of competition by Intel against the Alpha chip by Digital.

In response to this litigation, on August 12, 1997 Intel Corporation counterclaimed against Digital Corporation, accusing Digital of violating 14 Intel patents that may affect all aspects of Digital's business.

The outcome of this litigation is unclear and the legal and factual issues are quite complicated. However, the litigation between these two giant companies has the potential to go on for a long time and may threaten the fundamental business and next generation of growth for either company, should they lose the litigation.

This is an example of the high stakes in some patent cases today. It is important for players in all patentable industries to stake out their patent claims, as they develop new products and markets, for potential offensive or defensive action that may take place in later years to reallocate market share.

The Unocal Gasoline Patent

In the 1980's the State of California passed laws requiring less polluting gasoline. Unocal Corporation during that period filed a patent application for a type of gasoline that widely claims a range of gasoline formulations that comply with these requirements.

Presently, many of Unocal's gasoline competitors in the State of California claim that the Unocal patent improperly monopolizes any gasoline that is in compliance with the California statutes. In effect, the California pollution statute may require that only Unocal's patented gasoline may be sold, allege some of Unocal's complaining competitors.

There is a variety of lawsuits pending in California attacking the Unocal patent, by Unocal's competitors, including Mobil Oil Company, Atlantic Richfield Company, Exxon Corporation, Texaco Corporation, and Shell Corporation.

Some commentators estimate that if Unocal's patent is upheld, it may represent revenues of approximately $700,000,000 per year to Unocal in royalties for gasoline sold in the State of California, however, it is difficult to determine such possible revenues in advance.

It is interesting that the patent (if the patent is upheld and the pollution law not changed) may end up being a licensing bonanza for Unocal, even though Unocal no longer actually sells gasoline in California. (Unocal sold its gasoline sales operation since the issuance of the patent, but Unocal still owns the gasoline patent itself.)

The patent in question is U.S. Patent No. 5,288,393, for "gasoline fuel," filed December 13, 1990. The patent issued February 22, 1994. The patent has 155 claims, and states that "by controlling one or more properties of a gasoline fuel suitable for combustion in automobiles, the emissions of NO_x, CO, and/or hydrocarbons can be reduced".

Dell Computers and the Federal Trade Commission

The Unocal gas patent litigation may be reviewed in light of the Federal Trade Commission Decision and Order, Docket No. C-3658, May 20, 1996, regarding Dell Computers. Here the FTC ordered Dell to forego certain patent rights that Dell owned. Apparently, Dell had encouraged adoption of a technical standard by the Video Electronics Standards Association, and in the process stated that the standard should not violate any Dell patents. During the process, Dell had an unpublished U.S. patent application pending. After adoption and wide acceptance of the standard, the patent issued, and Dell apparently informed competitors that their use of the industry standard infringed Dell's patent rights. Dell may have had no intent to develop this situation, but the FTC found unreasonable restraint of competition, and ordered Dell to forego patent rights involved.

Haworth v. Steelcase

In another interesting patent case, over $200,000,000 and a permanent injunction was issued for the infringement of a furniture patent. See *Haworth, Inc. v. Steelcase*, 43 USPQ2d 1223 (1997). One interesting aspect of this case is that the patent in question is for a relatively low-tech item and illustrates the value of patents even in areas that are not usually considered to be exotic high technology areas. The Haworth patent enforced in this case is for movable wall panels for office cubicles, which contain extension cord-like wiring within the hollow spaces of the walls to connect to electric plugs installed in the walls. The plaintiff in this case received damages in excess of $200,000,000 for past infringement, and a permanent injunction against Steelcase for future infringement for the life of the patent. Estimates by some commentators of the size of this market in the U.S. exceeds $1,000,000,000 a year in sales.

Medical Procedures

In September 1996, 35 U.S.C. 287(c) was added to the U.S. Patent statute. This protects doctors, hospitals, and health maintenance organizations, universities, medical schools and others from enforcement of remedies for patent infringement (including injunctions, damages and attorneys' fees), for unauthorized performance of patented medical or surgical procedures. However, this does not protect the same parties from patent infringement remedies for commercial development and manufacturing, sale and distribution that constitutes infringement, nor does it protect other parties (such as manufacturers) from liability for patent infringement or inducement of infringement. Therefore, patent claims for medical devices and methods are still of great importance in the marketplace, especially since manufacturers are usually the target defendants for such infringement cases, and not individual doctors or hospitals.

Pending Legislation: Patent Prior Use Rights

Among several interesting pending patent bills in the U.S. Congress, is a bill that would invent something called "patent prior use rights." Basically, this proposal would create prior use rights for inventors who do not seek patents, as against a subsequent patentee, similar to the unregistered prior use rights now found in U.S. trademark law.

Under these proposed patent prior use rights, an inventor who does not seek patent protection could continue to use his invention despite later issuance of a patent to a later independent inventor. Under current U.S. law, an inventor who abandons his pursuit of a patent application or does not seek a patent application, could be subject to and actually infringe later patents obtained by a later independent inventor who diligently pursues patent protection.

This legislation for patent prior use rights currently is stalled in Congress, and it is unclear if it will pass. It is important to note that patent prior use rights are not part of the movement for global harmonization of patent law, and may be unique to the U.S. if passed. Furthermore, patent prior use rights are a new invention in the U.S. without any tradition.

Many commentators feel that patent prior use rights would be unconstitutional, if passed. The thought is that the Constitution in Article I, Section 8, authorizes Congress to give inventors exclusive rights for a limited period for their inventions. Of course, if patent prior use rights were given, they would be in addition to the patentee's rights. Consequently, the patentee's rights could not be exclusive because they would be shared with unpatented prior use rights. Hence, the prior use rights would be unconstitutional in that they would provide nonexclusive patent rights, which are specifically contrary to the congressional authority in the Constitution. That is, the U.S. Congress has no constitutional authority to create nonexclusive patent rights by prior use rights or otherwise.

On a larger level, prior use rights are directly contrary to the fundamental intent of the Constitution in providing for exclusive patent rights. The constitutional intent is to provide an inducement for inventors to come out of the world of trade secret, and to teach the world how to practice their invention. This is the point of the "not ... suppressed, or concealed it" language of 15 U.S.C. 102(g), and the reason that "known" in 15 U.S.C. 102(a) means "known to the public". In return for his disclosure to the public, the patentee obtains for a limited period an exclusive patent monopoly for his invention. The patent is superior to trade secrets during the patent term, because trade secrets protect the owner only as long as the secret remains a secret, and do not protect against subsequent independent inventors or certain types of disclosure.

The fundamental idea of patents is to undermine the old pre-constitutional practices of guilds and secret trade societies and instead "to promote science in the useful arts" by inducing all inventors to teach the world quickly about their inventions. Of course, if prior use rights were provided, this constitutional inducement to patent would be subverted, since any inventor would not risk losing rights to his invention to a subsequent patentee by remaining in the world of trade secrets. That is, if an inventor does not go to the Patent Office, but a subsequent inventor later does obtain a patent for the same invention, then the first inventor would still have the rights to use the invention under the new prior use rights.

Developments in this legislative area, if any, will be watched with interest.

If patent prior use rights are legislated in the U.S., then we can expect eventual litigation challenging their constitutionality. Of course, the outcome of this litigation would be unpredictable (as is the outcome of any litigation), even if policy and the Constitution would favor defeat of patent prior use rights.

Appendix 1

Fixing the Patent Statute: Title and Tort

The title (that is, ownership) section of the U.S. patent statute, 35 U.S.C. 261, is ambiguous and poorly written. As a result, title to U.S. patents, and the use of patents as collateral, is ambiguous. This undermines the value and usefulness of patents, and retards the growth and finance of technology businesses. No other type of property is so poorly treated by this congressional malpractice as is U.S. patent property.

The following bill is designed to correct these problems and to make patent title more secure, and transferable. Perfected security interests, reliable title searches, and title insurance for patents are also enabled. This will increase the predictability of transactions regarding patent assets, and reduce the costs of such transactions. This, in turn, will make patents worth more and facilitate the finance of technical innovation.

The approach of the bill follows well known approaches used for other types of property, such as real property, personal property, and specifically, airplanes.

The title provisions of the bill, Section 2 (a)–(f), are classic good government amendments that will be supported by the entire technology community. They will have no major opposition and not be controversial. The major lobbying effort must be educational, to increase the priority of the bill in the eyes of Congress.

The tort reform provision of the bill, Section 2 (g), can be expected to raise some controversy. It will unlock frozen technology owned by risk adverse inventions and institutions who are currently afraid to license their patents. Section 2 (g) can be separated out if this bill and passed separately.

Politically, the bill will promote economic and technologic development, at no cost to the taxpayer, and at microscopic cost to industry.

In summary, Section 2 of the bill provides the following approaches:

> Paragraph (a) enables creation of a perfected security interest in patent rights by filing a financing statement at the Patent and Trademark Office. Existing UCC and bankruptcy law applies to all subsequent issues, such as foreclosure procedure and priority among liens and creditors, once the security interest is created and perfected. This tracks the current approach for title to airplanes, in a statutory format.

Paragraph (b) tracks the existing statute.

> Paragraph (c) is new and enables written instruments of title to patent rights in addition to assignments.

> Paragraph (d) tracks the approach of the existing patent statute regarding use of acknowledgements.

Paragraph (e) requires that any instrument of patent title be recorded in writing in order to bind non-parties. This is the approach of the current statute towards assignments only. This is the core provision necessary for a complete title recording regime modeled after the real estate statutes and the UCC. This permits for the first time reliable patent title searches and reliable title opinions. In turn, this will enable patent title insurance.

Paragraph (f) provides for (i) an implied warranty of good and merchantable title, for any transaction in any interest in a patent or patent application, and (ii) an implied warranty that the patent interest is free of all liens and encumbrances.

Paragraph (g) protects from tort liability a mere inventor, patent holder, licensor, assignor, mortgagee or other individual in line for title for a patent. This is eliminates the chilling effect on risk adverse research institutions (such as universities, research hospitals and some large corporations), regarding licensing technology that they own but cannot commercialize.

The bill has been proposed to Congress. The approach has met with some interest in the House and Senate Judiciary Committees, and with some patent oriented trade groups. It is unclear what the bill's further development will be, and its support is encouraged.

I

104th CONGRESS
1st SESSION

S. ______________

To amend title 35, United States Code,
to make certain technical corrections to improve the
quality of title of United States patents.

IN THE SENATE

January ____, 1995

Mr. ______________ introduced the following bill.

A BILL

To amend title 35, United States Code,
to make certain technical corrections to improve the
quality of title to United States patents.

Be it enacted by the Senate and House of Representatives of the United States of America in Congress assembled,

SECTION 1. SHORT TITLE; REFERENCE

(a) SHORT TITLE.--This Act may be cited as the "Patents Technical Amendments Act of 1995."

(b) REFERENCE.--Except as expressly provided otherwise, whenever in this Act a section or other provision is amended or repealed, such amendment or repeal shall be considered to be made to that section or other provision of title 35, United States Code.

SEC. 2. PATENT TITLE.

Section 261 is amended by adding at the end the following: "§ 261. Patent Title.

"(a) Property. Subject to the provisions of this title, patents and patent applications shall have the attributes of general intangible personal property. A security interest in a patent or patent application is created by the execution of a financing statement referencing such patent or patent application as collateral, and such security interest is perfected by filing such financing statement in the Public Title Records of the U.S. Patent and Trademark Office. Nothing in this section shall pre-empt or amend existing state or federal law regarding the priority of security interests, the procedure to foreclose a security interest, the effect of such foreclosure, or the effect of a bankruptcy proceeding regarding a security interest.

"(b) Assignments. Patents, patent applications, and any interests therein, shall be assignable in law by an instrument in writing. The applicant, patentee, or his assigns or legal representatives may in like manner grant and convey an exclusive right under his application for patent, or patents, to the whole or any specified part of the United States.

"(c) Instruments of Title. Any interest in a patent or patent application, may be created by an instrument in writing by the owner of a greater and inclusive interest; however, this sentence shall not require a writing to create any such interest. Such instruments in writing may include, without limitation, licenses, sublicenses, mortgages, liens, security interests, options, contracts of purchase and sale, deeds of trust, wills, and written memoranda of selected provisions of agreements or written instruments creating or effecting interests in patents or patent applications. All such instruments in writing, and assignments and patents are referred to herein as "instruments of title."

"(d) Acknowledgements. A certificate of acknowledgement under the hand and official seal of a person authorized to administer oaths within the United States, or, in a foreign country, of a diplomatic or consular officer of the United States or an officer authorized to administer oaths whose authority is proved by a certificate of a diplomatic or consular officer of the United States, or apostille of an official designated by a foreign country which, by treaty or convention, accords like effect to apostilles of designated officials in the United States, shall be prima facie evidence of the execution of an assignment, grant or conveyance of a patent or application for patent.

"(e) Public Title Records. A prior assignment or instrument of title shall be void as against any subsequent assignment or instrument of title, unless the prior assignment or instrument of title is recorded in the Patent and Trademark Office within thirty (30) days after the date of execution or prior to the date of such recording of the subsequent assignment or instrument of title; however, nothing in this sentence requires such recording to make an assignment or instrument of title effective as to a party to that assignment or instrument of title, or as to any party with actual notice of such assignment or instrument title. Such filings may be made by facsimile, with the effective filing date being the date of receipt by the Patent and Trademark Office, if such filing is confirmed by mail or delivery within ten days of such facsimile. All recorded assignments and instruments of title shall be open to public inspection within thirty (30) days of such recording, where the subject property is a patent or published patent application, or where the owner of the subject property grants the public the right of inspection. The Office shall provide title searches upon request of those assignments and instruments of title that are open to public inspection, for such fees as the Office determines.

"(f) Implied Warranties of Title. Any seller, obligor, owner, or grantor of any interest in a patent or patent application in any transaction selling, obligating, encumbering, or granting any interest in a patent or patent application, shall be subject to the implied representations and warranties that (i) such party has good and merchantable title to such interest, (ii) such interest is free of all liens, encumbrances or defects, and (iii) such party is duly authorized and empowered to make such transaction. Such party may disclaim such implied representations and warranties in a writing delivered to the other parties in such transaction prior to or contemporaneously with such transaction.

"(g) Not Grounds for Liability In Tort. Inventing, owning, recording, collecting royalties for or otherwise benefiting from an invention, patent application or instrument of title (1) shall not be grounds for or a factor contributing to any liability for injury or damage arising from or connected to the manufacture, use or sale of that invention, patent application or instrument of title, (2) shall be deemed not to constitute or contribute to (i) negligence, contributory negligence, gross negligence, or intent to harm, or (ii) the design, manufacture, use, or sale of any product or service, (3) shall give arise to no duty to police or control the design, manufacture, use, or sale of any product or service arising from such invention, patent application, or instrument of title, and (4) except as otherwise provided herein, shall imply no representation or warranty regarding the invention, patent application, or instrument of title, or regarding products or services arising therefrom.

SEC. 3. CLERICAL AMENDMENT.

The item relating to section 261 in the table of sections at the beginning of chapter 26 is amended to read as follows:

"261 Patent Title."

SEC. 4. EFFECTIVE DATE AND APPLICABILITY

(a) IN GENERAL.--This Act and the amendments made by this Act shall take effect upon their passage into law.

(b) PROVISIONS SUPERSEDED.--On the effective date set forth in subsection (a), the provisions of section 261 of title 35, United States Code, are superseded with respect to all patents and patent applications containing one or more claims entitled to an effective filing date that is on or after such effective date.

Appendix 2

The U.S. Patent Statute, Title 35 U.S.C. (Summarized on the Head of a Pin, with Comments)

The invention of a "new and useful process, machine, manufacture or composition of matter" may obtain a patent. 35 U.S.C. 101. Software and engineered life forms are also patentable.

An invention may be barred from patenting by actions of the inventor (e.g., publication or sale before application), prior to invention by others, and other factors. 35 U.S.C. 102. Keep inventions confidential until patents are applied for. Do a patent search before using a new invention.

An invention must be not obvious at the time of conception, in order to be patentable, 35 U.S.C. 103, but new combinations of prior art components may be inventive and patentable. That is, new combinations of old parts may be patentable. The software analog to this statement is also true.

Generally, the U.S. follows a first to invent rule (not a first to file rule, like most of the world) to determine priority of claims among competing inventions. See generally 35 U.S.C. 102.

A U.S. patent gives the owner "the right to exclude others from making, using, offering for sale, or selling the invention throughout the U.S." or importing the invention [emphasis added], during the life of the patent. For patents resulting from applications filed on or after June 8, 1995, the patent term ends 20 years from the date of application in the U.S. (For patents filed before June 8, 1995, the term ends on the later of 17 years from the date of issue, or 20 years from the date of application, subject to any terminal disclaimers.) 35 U.S.C. 154. If the invention is a process, similar rights apply to products made by that process. 35 U.S.C. 271.

Patents cannot be renewed for an additional term, 35 U.S.C. 154, but new improvements of old inventions can be patented. 35 U.S.C. 101. Making the invention of one patent may infringe upon an earlier patent still in effect.

An invention must be diligently pursued and not abandoned, and generally kept confidential prior to patent application being filed, in order to maintain its patentability. 35 U.S.C. 102.

Patents may be assigned in writing only. 35 U.S.C. 261. Oral licenses are possible, but not advised. A joint owner of a patent may make the invention without the consent of or accounting to the other owners. 35 U.S.C. 262.

A patent owner may bring an action for infringement. Infringement may be direct, induced, or contributory. 35 U.S.C. 271. The court may permanently enjoin infringement, 35 U.S.C. 283, award compensatory damages, and treble those damages. 35 U.S.C. 284. An opinion of counsel regarding non-infringement prior to use of a new technical development can best avoid the possibility of treble damages. Remedies for non-commercial infringement of patents for medical procedures, against certain classes of infringers including doctors and hospitals, are limited. 35 U.S.C. 287 (c) (1).

There is a six year statute of limitations for infringement actions. 35 U.S.C. 286.

An issued patent is presumed valid. The burden of establishing invalidity of a patent rests on the party asserting invalidity. 35 U.S.C. 282.

A license that attempts to gain royalties from a patent beyond its term may constitute patent misuse and give grounds to any third party infringer to avoid enforcement of the patent.

To obtain patent rights outside the U.S., foreign applications must be filed in the targeted countries and regions. Foreign applications should be made prior to, simultaneously with, or within one year after, the U.S. application, depending on the specific facts. Other timing for the filing of foreign applications could result in the loss of foreign patent rights.

A patent is personal property. 35 U.S.C. 261. But assignments must be recorded at the U.S. Patent and Trademark Office to put third party buyers and mortgagees on notice. 35 U.S.C. 261. However, issues such as perfecting liens, foreclosures, post-foreclosure rights of redemption, and title searches are not clarified by statute.

Appendix 3

Patent and Trademark Office
United States Department of Commerce

Examination Guidelines for Computer-Related Inventions

Final Version

Appendix 3

Table of Contents

Appendix 3

Examination Guidelines for Computer-Related Inventions

I. Introduction

These *Examination Guidelines for Computer-Related Inventions*[1] ("Guidelines") are to assist Office personnel in the examination of applications drawn to computer-related inventions.[2] The Guidelines are based on the Office's current understanding of the law and are believed to be fully consistent with binding precedent of the Supreme Court, the Federal Circuit and the Federal Circuit's predecessor courts.

These Guidelines do not constitute substantive rulemaking and hence do not have the force and effect of law. These Guidelines have been designed to assist Office personnel in analyzing claimed subject matter for compliance with substantive law. Rejections will be based upon the substantive law and it is these rejections which are appealable. Consequently, any failure by Office personnel to follow the Guidelines is neither appealable nor petitionable.

The Guidelines alter the procedures Office personnel will follow when examining applications drawn to computer-related inventions and are equally applicable to claimed inventions implemented in either hardware or software. The Guidelines also clarify the Office's position on certain patentability standards related to this field of technology. Office personnel are to rely on these Guidelines in the event of any inconsistent treatment of issues between these Guidelines and any earlier provided guidance from the Office.

The Freeman-Walter-Abele[3] test may additionally be relied upon in analyzing claims directed solely to a process for solving a mathematical algorithm.

Office personnel have had difficulty in properly treating claims directed to methods of doing business. Claims should not be categorized as methods of doing business. Instead, such claims should be treated like any other process claims, pursuant to these Guidelines when relevant.[4]

The appendix includes a flow chart of the process Office personnel will follow in conducting examinations for computer-related inventions.

II. Determine What Applicant Has Invented and Is Seeking to Patent

It is essential that patent applicants obtain a prompt yet complete examination of their applications. Under the principles of compact prosecution, each claim should be reviewed for compliance with every statutory requirement for patentability in the initial review of the application, even if one or more claims are found to be deficient with respect to some statutory requirement. Thus, Office personnel should state all reasons and bases for rejecting claims in the first Office action. Deficiencies should be explained clearly, particularly when they serve as a basis for a rejection. Whenever practicable, Office personnel should indicate how rejections may be overcome and how problems may be resolved. A failure to follow this approach can lead to unnecessary delays in the prosecution of the application.

Prior to focusing on specific statutory requirements, Office personnel must begin examination by determining what, precisely, the applicant has invented and is seeking to patent,[5] and how the claims relate to and define that invention. Consequently, Office personnel will no longer begin examination by determining if a claim recites a "mathematical algorithm." Rather, they will review the complete specification, including the detailed description of the invention, any specific embodiments that have been disclosed, the claims and any specific utilities that have been asserted for the invention.

A. Identify and Understand Any Practical Application Asserted for the Invention

The subject matter sought to be patented must be a "useful" process, machine, manufacture or composition of matter, *i.e.*, it must have a practical application. The purpose of this requirement is to limit patent protection to inventions that possess a certain level of "real world" value, as opposed to subject matter that represents nothing more than an idea or concept, or is simply a starting point for future investigation or research.[6] Accordingly, a complete disclosure should contain some indication of the practical application for the claimed invention, *i.e.*, why the applicant believes the claimed invention is useful.

The utility of an invention must be within the "technological" arts.[7] A computer-related invention is within the technological arts. A practical application of a computer-related invention is statutory subject matter. This requirement can be discerned from the variously phrased prohibitions against the patenting of abstract ideas, laws of nature or natural phenomena. An invention that has a practical application in the technological arts satisfies the utility requirement.[8]

The applicant is in the best position to explain why an invention is believed useful. Office personnel should therefore focus their efforts on pointing out statements made in the specification that identify all practical applications for the invention. Office personnel should rely on such statements throughout the examination when assessing the invention for compliance with all statutory criteria. An applicant may assert more than one practical application, but only one is necessary to satisfy the utility requirement. Office personnel should review the entire disclosure to determine the features necessary to accomplish at least one asserted practical application.

B. Review the Detailed Disclosure and Specific Embodiments of the Invention to Determine What the Applicant Has Invented

The written description will provide the clearest explanation of the applicant's invention, by exemplifying the invention, explaining how it relates to the prior art and explaining the relative significance of various features of the invention. Accordingly, Office personnel should begin their evaluation of a computer-related invention as follows:

- determine what the programmed computer does when it performs the processes dictated by the software (*i.e.*, the *functionality* of the programmed computer);[9]

- determine how the computer is to be configured to provide that functionality (*i.e.*, what elements constitute the programmed computer and how those elements are configured and interrelated to provide the specified functionality); and

- if applicable, determine the *relationship* of the programmed computer to other subject matter outside the computer that constitutes the invention (*e.g.*, machines, devices, materials, or process steps other than those that are part of or performed by the programmed computer).[10]

Patent applicants can assist the Office by preparing applications that clearly set forth these aspects of a computer-related invention.

C. Review the Claims

The claims define the property rights provided by a patent, and thus require careful scrutiny. The goal of claim analysis is to identify the boundaries of the protection sought by the applicant and to understand how the claims relate to and define what the applicant has indicated is the invention. Office personnel must thoroughly analyze the language of a claim *before* determining if the claim complies with each statutory requirement for patentability.

Office personnel should begin claim analysis by identifying and evaluating each claim limitation. For processes, the claim limitations will define steps or acts to be performed. For products[11], the claim limitations will define discrete physical structures. The discrete physical structures may be comprised of hardware or a combination of hardware and software.

Office personnel are to correlate each claim limitation to all portions of the disclosure that describe the claim limitation. This is to be done in all cases, *i.e.*, whether or not the claimed invention is defined using means or step plus function language. The correlation step will ensure that Office personnel correctly interpret each claim limitation.

The subject matter of a properly construed claim is defined by the terms that limit its scope. It is this subject matter that must be examined. As a general matter, the grammar and intended meaning of terms used in a claim will dictate whether the language limits the claim scope. Language that *suggests or makes optional* but does not require steps to be performed or does not limit a claim to a particular structure does not limit the scope of a claim or claim limitation.[12]

Office personnel must rely on the applicant's disclosure to properly determine the meaning of terms used in the claims.[13] An applicant is entitled to be his or her own lexicographer, and in many instances will provide an explicit definition for certain terms used in the claims. Where an explicit definition is provided by the applicant for a term, that definition will control interpretation of the term as it is used in the claim. Office personnel should determine if the original disclosure provides a definition consistent with any assertions made by applicant.[14] If an applicant does not define a term in the specification, that term will be given its "common meaning."[15]

If the applicant asserts that a term has a meaning that conflicts with the term's art-accepted meaning, Office personnel should encourage the applicant to amend the claim to better reflect what applicant intends to claim as the invention. If the application becomes a patent, it becomes prior art against subsequent applications. Therefore, it is important for later search purposes to have the patentee employ commonly accepted terminology, particularly for searching text-searchable databases.

Office personnel must always remember to use the perspective of one of ordinary skill in the art. Claims and disclosures are not to be evaluated in a vacuum. If elements of an invention are well known in the art, the applicant does not have to provide a disclosure that describes those elements. In such a case the elements will be construed as encompassing any and every art-recognized hardware or combination of hardware and software technique for implementing the defined requisite functionalities.

Office personnel are to give claims their broadest reasonable interpretation in light of the supporting disclosure.[16] Where means plus function language is used to define the characteristics of a machine or manufacture invention, claim limitations must be interpreted to read on only the structures or materials disclosed in the specification and "equivalents thereof."[17] Disclosure may be express, implicit or inherent. Thus, at the outset, Office personnel must attempt to correlate claimed means to elements set forth in the written description. The written description includes the specification and the drawings. Office personnel are to give the claimed means plus function limitations their broadest reasonable interpretation consistent with all corresponding structures or materials described in the specification and their equivalents. Further guidance in interpreting the scope of equivalents is provided in the *Examination Guidelines For Claims Reciting A Means or Step Plus Function Limitation In Accordance With 35 U.S.C. 112, 6th Paragraph* ("Means Plus Function Guidelines").[18]

While it is appropriate to use the specification to determine what applicant intends a term to mean, a positive limitation from the specification cannot be read into a claim that does not impose that limitation. A broad interpretation of a claim by Office personnel will reduce the possibility that the claim, when issued, will be interpreted more broadly than is justified or intended. An applicant can always amend a claim during prosecution to better reflect the intended scope of the claim.

Finally, when evaluating the scope of a claim, *every* limitation in the claim must be considered.[19] Office personnel may not dissect a claimed invention into discrete elements and then evaluate the elements *in isolation*. Instead, the claim as a whole must be considered.

III. Conduct a Thorough Search of the Prior Art

Prior to classifying the claimed invention under § 101, Office personnel are expected to conduct a thorough search of the prior art. Generally, a thorough search involves reviewing both U.S. and foreign patents and non-patent literature. In many cases, the result of such a search will contribute to Office personnel's understanding of the invention. Both claimed and unclaimed aspects of the invention described in the specification should be searched if there is a reasonable expectation that the unclaimed aspects may be later claimed. A search must take into account any structure or material described in the specification and its equivalents which correspond to the claimed means plus function limitation, in accordance with 35 U.S.C. § 112, sixth paragraph and the Means Plus Function Guidelines.[20]

IV. Determine Whether the Claimed Invention Complies with 35 U.S.C. § 101

A. Consider the Breadth of 35 U.S.C. § 101 Under Controlling Law

As the Supreme Court has held, Congress chose the expansive language of § 101 so as to include "anything under the sun that is made by man."[21] Accordingly, § 101 of title 35, United States Code, provides:

> Whoever invents or discovers any new and useful process, machine, manufacture, or composition of matter, or any new and useful improvement thereof, may obtain a patent therefor, subject to the conditions and requirements of this title.[22]

As cast, § 101 defines four categories of inventions that Congress deemed to be the appropriate subject matter of a patent; namely, processes, machines, manufactures and compositions of matter. The latter three categories define "things" while the first category defines "actions" (*i.e.*, inventions that consist of a series of steps or acts to be performed).[23]

Federal courts have held that § 101 does have certain limits. First, the phrase "anything under the sun that is made by man" is limited by the text of § 101, meaning that one may only patent something that is a machine, manufacture, composition of matter or a process.[24] Second, § 101 requires that the subject matter sought to be patented be a "useful" invention. Accordingly, a complete definition of the scope of § 101, reflecting Congressional intent, is that any new and useful process, machine, manufacture or composition of matter under the sun that is made by man is the proper subject matter of a patent. Subject matter *not* within one of the four statutory invention categories or which is not "useful" in a patent sense is, accordingly, not eligible to be patented.

The subject matter courts have found to be outside the four statutory categories of invention is limited to abstract ideas, laws of nature and natural phenomena. While this is easily stated, determining whether an applicant is seeking to patent an abstract idea, a law of nature or a natural phenomenon has proven to be challenging. These three exclusions recognize that subject matter that is not a *practical application or use* of an idea, a law of nature or a natural phenomenon is not patentable.[25]

Courts have expressed a concern over "preemption" of ideas, laws of nature or natural phenomena.[26] The concern over preemption serves to bolster and justify the prohibition against the patenting of such subject matter. In fact, such concerns are only relevant to claiming a scientific truth or principle. Thus, a claim to an "abstract idea" is non-statutory because it does not represent a practical application of the idea, not because it would preempt the idea.

B. Classify the Claimed Invention as to Its Proper Statutory Category

To properly determine whether a claimed invention complies with the statutory invention requirements of § 101, Office personnel should classify each claim into *one or more* statutory or non-statutory categories. If the claim falls into a non-statutory category, that should not preclude complete examination of the application for satisfaction of all other conditions of patentability. This classification is *only an initial finding* at this point in the examination process that will be again assessed after the examination for compliance with §§ 102, 103 and 112 is completed and before issuance of any Office action on the merits.

If the invention as set forth in the written description is statutory, but the claims define subject matter that is not, the deficiency can be corrected by an appropriate amendment of the claims. In such a case, Office personnel should reject the claims drawn to non-statutory subject matter under § 101, but identify the features of the invention that would render the claimed subject matter statutory if recited in the claim.

1. Non-Statutory Subject Matter

Claims to computer-related inventions that are clearly non-statutory fall into the same general categories as non-statutory claims in other arts, namely natural phenomena such as magnetism, and abstract ideas or laws of nature which constitute "descriptive material." Descriptive material can be characterized as either "functional descriptive material" or "non-functional descriptive material." In this context, "functional descriptive material" consists of data structures[27] and computer programs which impart functionality when encoded on a computer-readable medium. "Non-functional descriptive material" includes but is not limited to music, literary works and a compilation or mere arrangement of data.

Both types of "descriptive material" are non-statutory when claimed as descriptive material *per se*. When functional descriptive material is recorded on some computer-readable medium it becomes structurally and functionally interrelated to the medium and will be statutory in most cases.[28] When non-functional descriptive material is recorded on some computer-readable medium, it is not structurally and functionally interrelated to the medium but is merely carried by the medium. Merely claiming non-functional descriptive material stored in a computer-readable medium does not make it statutory. Such a result would exalt form over substance.[29] Thus, non-statutory music does not become statutory by merely recording it on a compact disk. Protection for this type of work is provided under the copyright law.

Claims to processes that do nothing more than solve mathematical problems or manipulate abstract ideas or concepts are more complex to analyze and are addressed below. See sections IV.B.2(d) and IV.B.2(e).

(a) Functional Descriptive Material: "Data Structures" Representing Descriptive Material *Per Se* or Computer Programs Representing Computer Listings *Per Se*

Data structures not claimed as embodied in computer-readable media are descriptive material *per se* and are not statutory because they are neither physical "things" nor statutory processes.[30] Such claimed data structures do not define any structural and functional interrelationships between the data structure and other claimed aspects of the invention which permit the data structure's functionality to be realized. In contrast, a claimed computer-readable medium encoded with a data structure defines structural and functional interrelationships between the data structure and the medium which permit the data structure's functionality to be realized, and is thus statutory.

Similarly, computer programs claimed as computer listings *per se*, *i.e.*, the descriptions or expressions of the programs, are not physical "things," nor are they statutory processes, as they are not "acts" being performed. Such claimed computer programs do not define any structural and functional interrelationships between the computer program and other claimed aspects of the invention which permit the computer program's functionality to be realized. In contrast, a claimed computer-readable medium encoded with a computer program defines structural and functional interrelationships between the computer program and the medium which permit the computer program's functionality to be realized, and is thus statutory. Accordingly, it is important to distinguish claims that define descriptive material *per se* from claims that define statutory inventions.

Computer programs are often recited as part of a claim. Office personnel should determine whether the computer program is being claimed as part of an otherwise statutory manufacture or machine. In such a case, the claim remains statutory irrespective of the fact that a computer program is included in the claim. The same result occurs when a computer program is used in a computerized process where the computer executes the instructions set forth in the computer program. Only when the claimed invention taken as a whole is directed to a mere program listing, *i.e.*, to only its description or expression, is it descriptive material *per se* and hence non-statutory.

Since a computer program is merely a set of instructions capable of being executed by a computer, the computer program itself is not a process and Office personnel should treat a claim for a computer program, without the computer-readable medium needed to realize the computer program's functionality, as non-statutory functional descriptive material. When a computer program is claimed in a process where the computer is executing the computer program's instructions, Office personnel should treat the claim as a process claim. *See* Sections IV.B.2(b)-(e). When a computer program is recited in conjunction with a physical structure, such as a computer memory, Office personnel should treat the claim as a product claim. *See* Section IV.B.2(a).

(b) Non-Functional Descriptive Material

Descriptive material that cannot exhibit any functional interrelationship with the way in which computing processes are performed does not constitute a statutory process, machine, manufacture or composition of matter and should be rejected under § 101. Thus, Office personnel should consider the claimed invention as a whole to determine whether the necessary functional interrelationship is provided.

Where certain types of descriptive material, such as music, literature, art, photographs and mere arrangements or compilations of facts or data,[31] are merely stored so as to be read or outputted by a computer without creating any functional interrelationship, either as part of the stored data or as part of the computing processes performed by the computer, then such descriptive material alone does not impart functionality either to the data as so structured, or to the computer. Such "descriptive material" is not a process, machine, manufacture or composition of matter.

The policy that precludes the patenting of non-functional descriptive material would be easily frustrated if the same descriptive material could be patented when claimed as an article of manufacture.[32] For example, music is commonly sold to consumers in the format of a compact disc. In such cases, the known compact disc acts as nothing more than a carrier for non-functional descriptive material. The purely non-functional descriptive material cannot alone provide the practical application for the manufacture.

Office personnel should be prudent in applying the foregoing guidance. Non-functional descriptive material may be claimed in combination with other functional descriptive material on a computer-readable medium to provide the necessary functional and structural interrelationship to satisfy the requirements of § 101. The presence of the claimed non-functional descriptive material is not necessarily determinative of non-statutory subject matter. For example, a computer that recognizes a particular grouping of musical notes read from memory and upon recognizing that particular sequence, causes another defined series of notes to be played, defines a functional interrelationship among that data and the computing processes performed when utilizing that data, and as such is statutory because it implements a statutory process.

(c) Natural Phenomena Such as Electricity and Magnetism

Claims that recite nothing but the physical characteristics of a form of energy, such as a frequency, voltage, or the strength of a magnetic field, define energy or magnetism, *per se*, and as such are non-statutory natural phenomena.[33] However, a claim directed to a practical application of a natural phenomenon such as energy or magnetism is statutory.[34]

2. Statutory Subject Matter

(a) Statutory Product Claims[35]

If a claim defines a useful machine or manufacture by identifying the physical structure of the machine or manufacture in terms of its hardware or hardware and software combination, it defines a statutory product.[36]

A machine or manufacture claim may be one of two types: (1) a claim that encompasses any and every machine for performing the underlying process or any and every manufacture that can cause a computer to perform the underlying process, or (2) a claim that defines a specific machine or manufacture. When a claim is of the first type, Office personnel are to evaluate the underlying process the computer will perform in order to determine the patentability of the product.

(i) Claims that Encompass Any Machine or Manufacture Embodiment of a Process

Office personnel must treat each claim as a whole. The mere fact that a hardware element is recited in a claim does not necessarily limit the claim to a specific machine or manufacture.[37] If a product claim encompasses *any and every* computer implementation of a process, when read in light of the specification, it should be examined on the basis of the underlying process. Such a claim can be recognized as it will:

- define the physical characteristics of a computer or computer component exclusively as functions or steps to be performed on or by a computer, and

- encompass *any and every* product in the stated class (*e.g.*, computer, computer-readable memory) *configured in any manner* to perform that process.

Office personnel are reminded that finding a product claim to encompass any and every product embodiment of a process invention simply means that the Office will *presume* that the product claim encompasses any and every hardware or hardware platform and associated software implementation that performs the specified set of claimed functions. Because this is *interpretive* and *nothing more*, it does not provide any information as to the *patentability* of the applicant's underlying process or the product claim.

When Office personnel have reviewed the claim as a whole and found that it is not limited to a specific machine or manufacture, they shall identify how each claim limitation has been treated and set forth their reasons in support of their conclusion that the claim encompasses any and every machine or manufacture embodiment of a process. This will shift the burden to applicant to demonstrate why the claimed invention should be limited to a specific machine or manufacture.

If a claim is found to encompass any and every product embodiment of the underlying process, and if the underlying process is statutory, the product claim should be classified as a statutory product. By the same token, if the underlying process invention is found to be non-statutory, Office personnel should classify the "product" claim as a "non-statutory product." If the product claim is classified as being a non-statutory product on the basis of the underlying process, Office personnel should emphasize that they have considered all claim limitations and are basing their finding on the analysis of the underlying process.

(ii) Product Claims--Claims Directed to Specific Machines and Manufactures

If a product claim does not encompass any and every computer-implementation of a process, then it must be treated as a specific machine or manufacture. Claims that define a computer-related invention as a specific machine or specific article of manufacture must define the physical structure of the machine or manufacture in terms of its hardware or hardware and "specific software."[38] The applicant may define the physical structure of a programmed computer or its hardware or software components in any manner that can be clearly understood by a person skilled in the relevant art. Generally a claim drawn to a particular programmed computer should identify the elements of the computer and indicate how those elements are configured in either hardware or a combination of hardware and specific software.

To adequately define a specific computer memory, the claim must identify a general or specific memory and the specific software which provides the functionality stored in the memory.

A claim limited to a specific machine or manufacture, which has a practical application in the technological arts, is statutory. In most cases, a claim to a specific machine or manufacture will have a practical application in the technological arts.

(iii) Hypothetical Machine Claims Which Illustrate Claims of the Types Described in Sections IV.B.2(a)(i) and (ii)

Two applicants present a claim to the following process:

> A process for determining and displaying the structure of a chemical compound comprising:

(a) solving the wavefunction parameters for the compound to determine the structure of a compound; and

(b) displaying the structure of the compound determined in step (a).

Each applicant also presents a claim to the following apparatus:

> A computer system for determining the three dimensional structure of a chemical compound comprising:
>
> (a) means for determining the three dimensional structure of a compound; and
>
> (b) means for creating and displaying an image representing a three-dimensional perspective of the compound.

In addition, each applicant provides the noted disclosures to support the claims:

Disclosure: Applicant A

The disclosure describes specific software, *i.e.*, specific program code segments, that are to be employed to configure a general purpose microprocessor to create specific logic circuits. These circuits are indicated to be the "means" corresponding to the claimed means limitations.

Disclosure: Applicant B

The disclosure states that it would be a matter of routine skill to select an appropriate conventional computer system and implement the claimed process on that computer system. The disclosure does not have specific disclosure that corresponds to the two "means" limitations recited in the claim (*i.e.*, no specific software or logic circuit). The disclosure does have an explanation of how to solve the wavefunction equations of a chemical compound, and indicates that the solutions of those wavefunction equations can be employed to determine the physical structure of the corresponding compound.

Result: Applicant A

Claim defines specific computer, patentability stands independently from process claim.

Result: Applicant B

Claim encompasses any computer embodiment of process claim; patentability stands or falls with process claim.

Explanation: Applicant A	*Explanation: Applicant B*
Disclosure identifies the specific machine capable of performing the indicated functions.	Disclosure does not provide any information to distinguish the "implementation" of the process on a computer from the factors that will govern the patentability determination of the process *per se*. As such, the patentability of this apparatus claim will stand or fall with that of the process claim.

(b) Statutory Process Claims

A claim that requires one or more acts to be performed defines a process. However, not all processes are statutory under § 101. To be statutory, a claimed computer-related process must either: (1) result in a physical transformation outside the computer for which a practical application in the technological arts is either disclosed in the specification or would have been known to a skilled artisan (discussed in (i) below),[39] or (2) be limited by the language in the claim to a practical application within the technological arts (discussed in (ii) below).[40] The claimed practical application must be a further limitation upon the claimed subject matter if the process is confined to the internal operations of the computer. If a physical transformation occurs outside the computer, it is not necessary to claim the practical application. A disclosure that permits a skilled artisan to practice the claimed invention, *i.e.*, to put it to a practical use, is sufficient. On the other hand, it is necessary to claim the practical application if there is no physical transformation or if the process merely manipulates concepts or converts one set of numbers into another.

A claimed process is clearly statutory if it results in a physical transformation outside the computer, *i.e.*, falls into one or both of the following specific categories ("safe harbors").

(i) Safe Harbors

- Independent Physical Acts (Post-Computer Process Activity)

A process is statutory if it requires physical acts to be performed outside the computer independent of and following the steps to be performed by a programmed computer, where those acts involve the manipulation of tangible physical objects and result in the object having a different physical attribute or structure.[41] Thus, if a process claim includes one or more post-computer process steps that result in a physical transformation outside the computer (beyond merely conveying the direct result of the computer operation, *see* Section IV.B.2(d)(iii)), the claim is clearly statutory.

Examples of this type of statutory process include the following:

- A method of curing rubber in a mold which relies upon updating process parameters, using a computer processor to determine a time period for curing the rubber, using the computer processor to determine when the time period has been reached in the curing process and then opening the mold at that stage.

- A method of controlling a mechanical robot which relies upon storing data in a computer that represents various types of mechanical movements of the robot, using a computer processor to calculate positioning of the robot in relation to given tasks to be performed by the robot, and controlling the robot's movement and position based on the calculated position.

- **Manipulation of Data Representing Physical Objects or Activities (Pre-Computer Process Activity)**

Another statutory process is one that requires the measurements of physical objects or activities to be transformed outside of the computer into computer data,[42] where the data comprises signals corresponding to physical objects or activities external to the computer system, and where the process causes a physical transformation of the signals which are intangible representations of the physical objects or activities.[43]

Examples of this type of claimed statutory process include the following:

- A method of using a computer processor to analyze electrical signals and data representative of human cardiac activity by converting the signals to time segments, applying the time segments in reverse order to a high pass filter means, using the computer processor to determine the amplitude of the high pass filter's output, and using the computer processor to compare the value to a predetermined value. In this example the data is an intangible representation of physical activity, *i.e.*, human cardiac activity. The transformation occurs when heart activity is measured and an electrical signal is produced. This process has real world value in predicting vulnerability to ventricular tachycardia immediately after a heart attack.

- A method of using a computer processor to receive data representing Computerized Axial Tomography ("CAT") scan images of a patient, performing a calculation to determine the difference between a local value at a data point and an average value of the data in a region surrounding the point, and displaying the difference as a gray scale for each point in the image, and displaying the resulting image. In this example the data is an intangible representation of a physical object, *i.e.*, portions of the anatomy of a patient. The transformation occurs when the condition of the human body is measured with X-rays and the X-rays are converted into electrical digital signals that represent the condition of the human body. The real world value of the invention lies in creating a new CAT scan image of body tissue without the presence of bones.

- A method of using a computer processor to conduct seismic exploration, by imparting spherical seismic energy waves into the earth from a seismic source, generating a plurality of reflected signals in response to the seismic energy waves at a set of receiver positions in an array, and summing the reflection signals to produce a signal simulating the reflection response of the earth to the seismic energy. In this example, the electrical signals processed by the computer represent reflected seismic energy. The transformation occurs by converting the spherical seismic energy waves into electrical signals which provide a geophysical representation of formations below the earth's surface. Geophysical exploration of formations below the surface of the earth has real world value.

If a claim does not clearly fall into one or both of the safe harbors, the claim may still be statutory if it is limited by the language in the claim to a practical application in the technological arts.

(ii) Computer-Related Processes Limited to a Practical Application in the Technological Arts

There is always some form of physical transformation within a computer because a computer acts on signals and transforms them during its operation and changes the state of its components during the execution of a process. Even though such a physical transformation occurs within a computer, such activity is not determinative of whether the process is statutory because such transformation alone does not distinguish a statutory computer process from a non-statutory computer process. What is determinative is not how the computer performs the process, but what the computer does to achieve a practical application.[44]

A process that merely manipulates an abstract idea or performs a purely mathematical algorithm is non-statutory despite the fact that it might inherently have some usefulness.[45] For such subject matter to be statutory, the claimed process must be limited to a practical application of the abstract idea or mathematical algorithm in the technological arts.[46] For example, a computer process that simply calculates a mathematical algorithm that models noise is non-statutory. However, a claimed process for digitally filtering noise employing the mathematical algorithm is statutory.

Examples of this type of claimed statutory process include the following:

- A computerized method of optimally controlling transfer, storage and retrieval of data between cache and hard disk storage devices such that the most frequently used data is readily available.

- A method of controlling parallel processors to accomplish multi-tasking of several computing tasks to maximize computing efficiency.[47]

- A method of making a word processor by storing an executable word processing application program in a general purpose digital computer's memory, and executing the stored program to impart word processing functionality to the general purpose digital computer by changing the state of the computer's arithmetic logic unit when program instructions of the word processing program are executed.

- A digital filtering process for removing noise from a digital signal comprising the steps of calculating a mathematical algorithm to produce a correction signal and subtracting the correction signal from the digital signal to remove the noise.

(c) Non-Statutory Process Claims

If the "acts" of a claimed process manipulate only numbers, abstract concepts or ideas, or signals representing any of the foregoing, the acts are not being applied to appropriate subject matter. Thus, a process consisting solely of mathematical operations, *i.e.*, converting one set of numbers into another set of numbers, does not manipulate appropriate subject matter and thus cannot constitute a statutory process.

In practical terms, claims define non-statutory processes if they:

- consist solely of mathematical operations without some claimed practical application (*i.e.*, executing a "mathematical algorithm"); or

- simply manipulate abstract ideas, *e.g.*, a bid[48] or a bubble hierarchy,[49] without some claimed practical application.

A claimed process that consists solely of mathematical operations is non-statutory whether or not it is performed on a computer. Courts have recognized a distinction between types of mathematical algorithms, namely, some define a "law of nature" in mathematical terms and others merely describe an "abstract idea."[50]

Certain mathematical algorithms have been held to be non-statutory because they represent a mathematical definition of a law of nature or a natural phenomenon. For example, a mathematical algorithm representing the formula $E=mc^2$ is a "law of nature"--it defines a "fundamental scientific truth" (*i.e.*, the relationship between energy and mass). To comprehend how the law of nature relates to any object, one invariably has to perform certain steps (*e.g.*, multiplying a number representing the mass of an object by the square of a number representing the speed of light). In such a case, a claimed process which consists solely of the steps that one must follow to solve the mathematical representation of $E=mc^2$ is indistinguishable from the law of nature and would "preempt" the law of nature. A patent cannot be granted on such a process.

Other mathematical algorithms have been held to be non-statutory because they merely describe an abstract idea. An "abstract idea" may simply be *any* sequence of mathematical operations that are combined to solve a mathematical problem. The concern addressed by holding such subject matter non-statutory is that the mathematical operations merely describe an idea and do not define a process that represents a practical application of the idea.

Accordingly, when a claim reciting a mathematical algorithm is found to define non-statutory subject matter the basis of the § 101 rejection must be that, when taken as a whole, the claim recites a law of nature, a natural phenomenon, or an abstract idea.

(d) Certain Claim Language Related to Mathematical Operation Steps of a Process

(i) Intended Use or Field of Use Statements

Claim language that simply specifies an intended use or field of use for the invention generally will not limit the scope of a claim, particularly when only presented in the claim preamble. Thus, Office personnel should be careful to properly interpret such language.[51] When such language is treated as non-limiting, Office personnel should expressly identify in the Office action the claim language that constitutes the intended use or field of use statements and provide the basis for their findings. This will shift the burden to applicant to demonstrate why the language is to be treated as a claim limitation.

(ii) Necessary Antecedent Step to Performance of a Mathematical Operation or Independent Limitation on a Claimed Process

In some situations, certain acts of "collecting" or "selecting" data for use in a process consisting of one or more mathematical operations will not further limit a claim beyond the specified mathematical operation step(s). Such acts merely determine values for the variables used in the mathematical formulae used in making the calculations.[52] In other words, the acts are dictated by nothing other than the performance of a mathematical operation.[53]

If a claim requires acts to be performed to *create* data that will then be used in a process representing a *practical application* of one or more mathematical operations, those acts *must* be treated as further limiting the claim beyond the mathematical operation(s) *per se*. Such acts are data gathering steps not dictated by the algorithm but by other limitations which require certain antecedent steps and as such constitute an independent limitation on the claim.

Examples of acts that independently limit a claimed process involving mathematical operations include:

- a method of conducting seismic exploration which requires generating and manipulating signals from seismic energy waves *before* "summing" the values represented by the signals;[54] and

- a method of displaying X-ray attenuation data as a signed gray scale signal in a "field" using a particular algorithm, where the antecedent steps require generating the data using a particular machine (*e.g.*, a computer tomography scanner).[55]

Examples of steps that do not independently limit one or more mathematical operation steps include:

- "perturbing" the values of a set of process inputs, where the subject matter "perturbed" was a number and the act of "perturbing" consists of substituting the numerical values of variables;[56] and

- selecting a set of arbitrary measurement point values.[57]

Such steps do not impose independent limitations on the scope of the claim beyond those required by the mathematical operation limitation.

(iii) Post-Mathematical Operation Step Using Solution or Merely Conveying Result of Operation

In some instances, certain kinds of post-solution "acts" will not further limit a process claim beyond the performance of the preceding mathematical operation step even if the acts are recited in the body of a claim. If, however, the claimed acts represent some "significant use" of the solution, those acts will invariably impose an independent limitation on the claim. A "significant use" is any activity which is more than merely outputting the direct result of the mathematical operation. Office personnel are reminded to rely on the applicant's characterization of the significance of the acts being assessed to resolve questions related to their relationship to the mathematical operations recited in the claim and the invention as a whole.[58] Thus, if a claim requires that the direct result of a mathematical operation be evaluated and transformed into something else, Office personnel cannot treat the subsequent steps as being *indistinguishable* from the performance of the mathematical operation and thus not further limiting on the claim. For example, acts that require the conversion of a series of numbers representing values of a wavefunction equation for a chemical compound into values representing an image that conveys information about the three-dimensional structure of the compound and the displaying of the three-dimensional structure cannot be treated as being part of the mathematical operations.

Office personnel should be especially careful when reviewing claim language that requires the performance of "post-solution" steps to ensure that claim limitations are not ignored.

Examples of steps found not to independently limit a process involving one or more mathematical operation steps include:

- step of "updating alarm limits" found to constitute changing the number value of a variable to represent the result of the calculation;[59]

- final step of magnetically recording the result of a calculation;[60]

- final step of "equating" the process outputs to the values of the last set of process inputs found to constitute storing the result of calculations;[61]

- final step of displaying result of a calculation "as a shade of gray rather than as simply a number" found to not constitute distinct step where the data were numerical values that did not represent anything;[62] and

- step of "transmitting electrical signals representing" the result of calculations.[63]

(e) Manipulation of Abstract Ideas Without a Claimed Practical Application

A process that consists solely of the manipulation of an abstract idea without any limitation to a practical application is non-statutory.[64] Office personnel have the burden to establish a *prima facie* case that the claimed invention taken as a whole is directed to the manipulation of abstract ideas without a practical application.

In order to determine whether the claim is limited to a practical application of an abstract idea, Office personnel must analyze the claim as a whole, in light of the specification, to understand what subject matter is being manipulated and how it is being manipulated. During this procedure, Office personnel must evaluate any statements of intended use or field of use, any data gathering step and any post-manipulation activity. *See* section IV.B.2(d) above for how to treat various types of claim language. Only when the claim is devoid of any limitation to a practical application in the technological arts should it be rejected under § 101. Further, when such a rejection is made, Office personnel must expressly state how the language of the claims has been interpreted to support the rejection.

V. Evaluate Application for Compliance with 35 U.S.C. § 112

Office personnel should begin their evaluation of an application's compliance with § 112 by considering the requirements of § 112, second paragraph. The second paragraph contains two separate and distinct requirements: (1) that the claim(s) set forth the subject matter applicants regard as the invention, and (2) that the claim(s) particularly point out and distinctly claim the invention. An application will be deficient under § 112, second paragraph when (1) evidence including admissions, other than in the application as filed, shows applicant has stated that he or she regards the invention to be different from what is claimed, or when (2) the scope of the claims is unclear.

After evaluation of the application for compliance with § 112, second paragraph, Office personnel should then evaluate the application for compliance with the requirements of § 112, first paragraph. The first paragraph contains three separate and distinct requirements: (1) adequate written description, (2) enablement, and (3) best mode. An application will be deficient under § 112, first paragraph when the written description is not adequate to identify what the applicant has invented, or when the disclosure does not enable one skilled in the art to make and use the invention as claimed without undue experimentation. Deficiencies related to disclosure of the best mode for carrying out the claimed invention are not usually encountered during examination of an application because evidence to support such a deficiency is seldom in the record.

If deficiencies are discovered with respect to § 112, Office personnel must be careful to apply the appropriate paragraph of § 112.

A. Determine Whether the Claimed Invention Complies with 35 U.S.C. § 112, Second Paragraph Requirements

1. Claims Setting Forth the Subject Matter Applicant Regards as Invention

Applicant's specification must conclude with claim(s) that set forth the subject matter which the applicant regards as the invention. The invention set forth in the claims is presumed to be that which applicant regards as the invention, unless applicant considers the invention to be something different from what has been claimed as shown by evidence, including admissions, outside the application as filed. An applicant may change what he or she regards as the invention during the prosecution of the application.

2. Claims Particularly Pointing Out and Distinctly Claiming the Invention

Office personnel shall determine whether the claims set out and circumscribe the invention with a reasonable degree of precision and particularity. In this regard, the definiteness of the language must be analyzed, not in a vacuum, but always in light of the teachings of the disclosure as it would be interpreted by one of ordinary skill in the art. Applicant's claims, interpreted in light of the disclosure, must reasonably apprise a person of ordinary skill in the art of the invention. However, the applicant need not explicitly recite in the claims every feature of the invention. For example, if an applicant indicates that the invention is a particular computer, the claims do not have to recite every element or feature of the computer. In fact, it is preferable for claims to be drafted in a form that emphasizes what the applicant has invented (*i.e.*, what is new rather than old).

A means plus function limitation is distinctly claimed if the description makes it clear that the means corresponds to well-defined structure of a computer or computer component implemented in either hardware or software and its associated hardware platform. Such means may be defined as:

- a programmed computer with a particular functionality implemented in hardware or hardware and software;

- a logic circuit or other component of a programmed computer that performs a series of specifically identified operations dictated by a computer program; or

- a computer memory encoded with executable instructions representing a computer program that can cause a computer to function in a particular fashion.

The scope of a "means" limitation is defined as the corresponding structure or material (*e.g.*, a specific logic circuit) set forth in the written description and equivalents.[65] Thus, a claim using means plus function limitations without corresponding disclosure of specific structures or materials that are not well-known fails to particularly point out and distinctly claim the invention. For example, if the applicant discloses only the functions to be performed and provides no express, implied or inherent disclosure of hardware or a combination of hardware and software that performs the functions, the application has not disclosed any "structure" which corresponds to the claimed means. Office personnel should reject such claims under § 112, second paragraph. The rejection shifts the burden to the applicant to describe at least one specific structure or material that corresponds to the claimed means in question, and to identify the precise location or locations in the specification where a description of at least one embodiment of that claimed means can be found. In contrast, if the corresponding structure is disclosed to be a memory or logic circuit that has been configured in some manner to perform that function (*e.g.*, using a defined computer program), the application has disclosed "structure" which corresponds to the claimed means.

When a claim or part of a claim is defined in computer program code, whether in source or object code format, a person of skill in the art must be able to ascertain the metes and bounds of the claimed invention. In certain circumstances, as where self-documenting programming code is employed, use of programming language in a claim would be permissible because such program source code presents "sufficiently high-level language and descriptive identifiers" to make it universally understood to others in the art without the programmer having to insert any comments.[66] Applicants should be encouraged to functionally define the steps the computer will perform rather than simply reciting source or object code instructions.

B. Determine Whether the Claimed Invention Complies with 35 U.S.C. § 112, First Paragraph Requirements

1. Adequate Written Description

The satisfaction of the enablement requirement does not satisfy the written description requirement.[67] For the written description requirement, an applicant's specification must reasonably convey to those skilled in the art that the applicant was in possession of the claimed invention as of the date of invention. The claimed invention subject matter need not be described literally, *i.e.*, using the same terms, in order for the disclosure to satisfy the description requirement.

2. Enabling Disclosure

An applicant's specification must enable a person skilled in the art to make and use the claimed invention without undue experimentation. The fact that experimentation is complex, however, will not make it undue if a person of skill in the art typically engages in such complex experimentation. For a computer-related invention, the disclosure must enable a skilled artisan to configure the computer to possess the requisite functionality, and, where applicable, interrelate the computer with other elements to yield the claimed invention, without the exercise of undue experimentation. The specification should disclose *how* to configure a computer to possess the requisite functionality or *how* to integrate the programmed computer with other elements of the invention, unless a skilled artisan would know how to do so without such disclosure.[68]

For many computer-related inventions, it is not unusual for the claimed invention to involve more than one field of technology. For such inventions, the disclosure must satisfy the enablement standard for each aspect of the invention.[69] As such, the disclosure must teach a person skilled in each art how to make and use the relevant aspect of the invention without undue experimentation. For example, to enable a claim to a programmed computer that determines and displays the three-dimensional structure of a chemical compound, the disclosure must

- enable a person skilled in the art of molecular modeling to understand and practice the underlying molecular modeling processes; and

- enable a person skilled in the art of computer programming to create a program that directs a computer to create and display the image representing the three-dimensional structure of the compound.

In other words, the disclosure corresponding to each aspect of the invention must be enabling to a person skilled in each respective art.

In many instances, an applicant will describe a programmed computer by outlining the significant elements of the programmed computer using a functional block diagram. Office personnel should review the specification to ensure that along with the functional block diagram the disclosure provides information that adequately describes each "element" in hardware or hardware and its associated software and how such elements are interrelated.[70]

VI. Determine Whether the Claimed Invention Complies with 35 U.S.C. §§ 102 and 103

As is the case for inventions in any field of technology, assessment of a claimed computer-related invention for compliance with § 102 and § 103 begins with a comparison of the claimed subject matter to what is known in the prior art. If no differences are found between the claimed invention and the prior art, the claimed invention lacks novelty and is to be rejected by Office personnel under § 102. Once distinctions are identified between the claimed invention and the prior art, those distinctions must be assessed and resolved in light of the knowledge possessed by a person of ordinary skill in the art. Against this backdrop, one must determine whether the invention would have been obvious at the time the invention was made. If not, the claimed invention satisfies § 103. Factors and considerations dictated by law governing § 103 apply without modification to computer-related inventions.

If the difference between the prior art and the claimed invention is limited to descriptive material stored on or employed by a machine, Office personnel must determine whether the descriptive material is functional descriptive material or non-functional descriptive material, as described supra in Section IV. Functional descriptive material is a limitation in the claim and must be considered and addressed in assessing patentability under § 103. Thus, a rejection of the claim as a whole under § 103 is inappropriate unless the functional descriptive material would have been suggested by the prior art. Non-functional descriptive material cannot render non-obvious an invention that would have otherwise been obvious.[71]

Common situations involving non-functional descriptive material are:

- a computer-readable storage medium that differs from the prior art solely with respect to non-functional descriptive material, such as music or a literary work, encoded on the medium,

- a computer that differs from the prior art solely with respect to non-functional descriptive material that cannot alter how the machine functions (*i.e.*, the descriptive material does not reconfigure the computer), or

- a process that differs from the prior art only with respect to non-functional descriptive material that cannot alter *how* the process steps are to be performed to achieve the utility of the invention.

Thus, if the prior art suggests storing a song on a disk, merely choosing a *particular* song to store on the disk would be presumed to be well within the level of ordinary skill in the art at the time the invention was made. The difference between the prior art and the claimed invention is simply a rearrangement of non-functional descriptive material.

VII. Clearly Communicate Findings, Conclusions and Their Bases

Once Office personnel have concluded the above analyses of the claimed invention under all the statutory provisions, including §§ 101, 112, 102 and 103, they should review all the proposed rejections and their bases to confirm their correctness. Only then should any rejection be imposed in an Office action. The Office action should clearly communicate the findings, conclusions and reasons which support them.

Footnotes to Guidelines

[1] These Guidelines are final and replace the *Proposed Examination Guidelines for Computer-Implemented Inventions*, 60 Fed. Reg. 28,778 (June 2, 1995) and the supporting legal analysis issued on October 3, 1995.

[2] "Computer-related inventions" include inventions implemented in a computer and inventions employing computer-readable media.

[3] *In re Abele*, 684 F.2d 902, 905-07, 214 USPQ 682, 685-87 (CCPA 1982); *In re Walter*, 618 F.2d 758, 767, 205 USPQ 397, 406-07 (CCPA 1980); *In re Freeman*, 573 F.2d 1237, 1245, 197 USPQ 464, 471 (CCPA 1978).

[4] *See, e.g., In re Toma*, 575 F.2d 872, 877-78, 197 USPQ 852, 857 (CCPA 1978); *In re Musgrave*, 431 F.2d 882, 893, 167 USPQ 280, 289-90 (CCPA 1970). *See also In re Schrader*, 22 F.3d 290, 297-98, 30 USPQ2d 1455, 1461-62 (Fed. Cir. 1994) (Newman, J., dissenting); *Paine, Webber, Jackson & Curtis, Inc. v. Merrill Lynch, Pierce, Fenner & Smith, Inc.*, 564 F. Supp. 1358, 1368-69, 218 USPQ 212, 220 (D. Del. 1983).

[5] As the courts have repeatedly reminded the Office: "The goal is to answer the question "'What did applicants invent?'" *Abele*, 684 F.2d at 907, 214 USPQ at 687. *Accord, e.g., Arrhyth-*

mia Research Tech. v. Corazonix Corp., 958 F.2d 1053, 1059, 22 USPQ2d 1033, 1038 (Fed. Cir. 1992).

[6] *Brenner v. Manson*, 383 U.S. 519, 528-36, 148 USPQ 689, 693-96 (1966); *In re Ziegler*, 992 F.2d 1197, 1200-03, 26 USPQ2d 1600, 1603-06 (Fed. Cir. 1993).

[7] *See, e.g., Musgrave*, 431 F.2d at 893, 167 USPQ at 289-90, *cited with approval in Schrader*, 22 F.3d at 297, 30 USPQ2d at 1461 (Newman, J., dissenting). The definition of "technology" is the "application of science and engineering to the development of machines and procedures in order to enhance or improve human conditions, or at least to improve human efficiency in some respect." Computer Dictionary 384 (Microsoft Press, 2d ed. 1994).

[8] *E.g., In re Alappat*, 33 F.3d 1526, 1543, 31 USPQ2d 1545, 1556-57 (Fed. Cir. 1994) (in banc) (quoting *Diamond v. Diehr*, 450 U.S. 175, 192, 209 USPQ 1, 10 (1981)). *See also id.* at 1569, 31 USPQ2d at 1578-79 (Newman, J., concurring) ("unpatentability of the principle does not defeat patentability of its practical applications") (citing *O'Reilly v. Morse*, 56 U.S. (15 How.) 62, 114-19 (1854)); *Arrhythmia*, 958 F.2d at 1056, 22 USPQ2d at 1036; *Musgrave*, 431 F.2d at 893, 167 USPQ at 289-90 ("All that is necessary, in our view, to make a sequence of operational steps a statutory 'process' within 35 U.S.C. 101 is that it be in the technological arts so as to be in consonance with the Constitutional purpose to promote the progress of 'useful arts.' Const. Art. 1, sec. 8.").

[9] *Arrhythmia*, 958 F.2d at 1057, 22 USPQ2d at 1036:

> It is of course true that a modern digital computer manipulates data, usually in binary form, by performing mathematical operations, such as addition, subtraction,

multiplication, division, or bit shifting, on the data. But this is only *how* the computer does what it does. Of importance is the significance of the data and their manipulation in the real world, *i.e.*, *what* the computer is doing.

[10] Many computer-related inventions do not consist solely of a computer. Thus, Office personnel should identify those claimed elements of the computer-related invention that are not part of the programmed computer, and determine how those elements relate to the programmed computer. Office personnel should look for specific information that explains the role of the programmed computer in the overall process or machine and how the programmed computer is to be integrated with the other elements of the apparatus or used in the process.

[11] Products may be either machines, manufactures or compositions of matter. Product claims are claims that are directed to either machines, manufactures or compositions of matter.

[12] Examples of language that *may* raise a question as to the limiting effect of the language in a claim:

(a) statements of intended use or field of use,
(b) "adapted to" or "adapted for" clauses,
(c) "wherein" clauses, or
(d) "whereby" clauses.

This list of examples is not intended to be exhaustive.

[13] *Markman v. Westview Instruments*, 52 F.3d 967, 980, 34 USPQ2d 1321, 1330 (Fed. Cir.) (in banc), *cert. granted*, 116 S. Ct. 40 (1995).

[14] *See, e.g.*, *In re Paulsen*, 30 F.3d 1475, 1480, 31 USPQ2d 1671, 1674 (Fed. Cir. 1994) (inventor may define specific terms used to describe invention, but must do so "with reasonable clarity, deliberateness, and precision" and, if done, must "'set out his uncommon definition in some manner within the patent disclosure' so as to give one of ordinary skill in the art notice of the change" in meaning) (quoting *Intellicall, Inc. v. Phonometrics, Inc.*, 952 F.2d 1384, 1387-88, 21 USPQ2d 1383, 1386 (Fed. Cir. 1992)).

[15] *Id.* at 1480, 31 USPQ2d at 1674.

[16] *See, e.g.*, *In re Zletz*, 893 F.2d 319, 321-22, 13 USPQ2d 1320, 1322 (Fed. Cir. 1989) ("During patent examination the pending claims must be interpreted as broadly as their terms reasonably allow. . . . The reason is simply that during patent prosecution when claims can be amended, ambiguities should be recognized, scope and breadth of language explored, and clarification imposed. . . . An essential purpose of patent examination is to fashion claims that are precise, clear, correct, and unambiguous. Only in this way can uncertainties of claim scope be removed, as much as possible, during the administrative process.").

[17] Two *in banc* decisions of the Federal Circuit have made clear that the Office is to interpret means plus function language according to 35 U.S.C. § 112, sixth paragraph. In the first, *In re Donaldson*, 16 F.3d 1189, 1193, 29 USPQ2d 1845, 1848 (Fed. Cir. 1994), the court held:

> The plain and unambiguous meaning of paragraph six is that one construing means-plus-function language in a claim must look to the specification and interpret that language in light of the corresponding structure, material, or acts described therein, and equivalents thereof, to the

extent that the specification provides such disclosure. Paragraph six does not state or even suggest that the PTO is exempt from this mandate, and there is no legislative history indicating that Congress intended that the PTO should be. Thus, this court must accept the plain and precise language of paragraph six.

Consistent with *Donaldson*, in the second decision, *Alappat*, 33 F.3d at 1540, 31 USPQ2d at 1554, the Federal Circuit held:

> Given Alappat's disclosure, it was error for the Board majority to interpret each of the means clauses in claim 15 so broadly as to "read on any and every means for performing the function" recited, as it said it was doing, and then to conclude that claim 15 is nothing more than a process claim wherein each means clause represents a step in that process. Contrary to suggestions by the Commissioner, this court's precedents do not support the Board's view that the particular apparatus claims at issue in this case may be viewed as nothing more than process claims.

[18] 1162 O.G. 59 (May 17, 1994).

[19] *See, e.g., Diamond v. Diehr*, 450 U.S. at 188-89, 209 USPQ at 9 ("In determining the eligibility of respondents' claimed process for patent protection under § 101, their claims must be considered as a whole. It is inappropriate to dissect the claims into old and new elements and then to ignore the presence of the old elements in the analysis. This is particularly true in a process claim because a new combination of steps in a process may be patentable even though all the constituents of the combination were well known and in common use before the combination was made.").

[20] *See supra* note 18 and accompanying text.

[21] *Diamond v. Chakrabarty*, 447 U.S. 303, 308-09, 206 USPQ 193, 197 (1980):

> In choosing such expansive terms as "manufacture" and "composition of matter," modified by the comprehensive "any," Congress plainly contemplated that the patent laws would be given wide scope. The relevant legislative history also supports a broad construction. The Patent Act of 1793, authored by Thomas Jefferson, defined statutory subject matter as "any new and useful art, machine, manufacture, or composition of matter, or any new or useful improvement [thereof]." Act of Feb. 21, 1793, § 1, 1 Stat. 319. The Act embodied Jefferson's philosophy that "ingenuity should receive a liberal encouragement." 5 Writings of Thomas Jefferson 75-76 (Washington ed. 1871). See *Graham v. John Deere Co.*, 383 U.S. 1, 7-10 (1966). Subsequent patent statutes in 1836, 1870, and 1874 employed this same broad language. In 1952, when the patent laws were recodified, Congress replaced the word "art" with "process," but otherwise left Jefferson's language intact. The Committee Reports accompanying the 1952 Act inform us that Congress intended statutory subject matter to "include anything under the sun that is made by man." S. Rep. No. 1979, 82d Cong., 2d Sess. 5 (1952); H.R. Rep. No. 1923, 82d Cong., 2d Sess. 6 (1952).

This perspective has been embraced by the Federal Circuit:

> The plain and unambiguous meaning of § 101 is that any new and useful process, machine, manufacture, or composition of matter, or any new and useful improve-

ment thereof, may be patented if it meets the requirements for patentability set forth in Title 35, such as those found in §§ 102, 103, and 112. The use of the expansive term "any" in § 101 represents Congress's intent not to place any restrictions on the subject matter for which a patent may be obtained beyond those specifically recited in § 101 and the other parts of Title 35. . . . Thus, it is improper to read into § 101 limitations as to the subject matter that may be patented where the legislative history does not indicate that Congress clearly intended such limitations. [*Alappat*, 33 F.3d at 1542, 31 USPQ2d at 1556.]

[22] 35 U.S.C. § 101 (1994).

[23] *See* 35 U.S.C. § 100(b) ("The term 'process' means process, art, or method, and includes a new use of a known process, machine, manufacture, composition of matter, or material.").

[24] *E.g.*, *Alappat*, 33 F.3d at 1542, 31 USPQ2d at 1556; *In re Warmerdam*, 33 F.3d 1354, 1358, 31 USPQ2d 1754, 1757 (Fed. Cir. 1994).

[25] *See, e.g.*, *Rubber-Tip Pencil Co. v. Howard*, 87 U.S. 498, 507 (1874) ("idea of itself is not patentable, but a new device by which it may be made practically useful is"); *Mackay Radio & Telegraph Co. v. Radio Corp. of America*, 306 U.S. 86, 94 (1939) ("While a scientific truth, or the mathematical expression of it, is not patentable invention, a novel and useful structure created with the aid of knowledge of scientific truth may be."); *Warmerdam*, 33 F.3d at 1360, 31 USPQ2d at 1759 ("steps of 'locating' a medial axis, and 'creating' a bubble hierarchy . . . describe nothing more than the manipulation of basic mathematical constructs, the paradigmatic 'abstract idea'").

[26] The concern over preemption was expressed as early as 1852. *See Le Roy v. Tatham*, 55 U.S. 156, 175 (1852) ("A principle, in the abstract, is a fundamental truth; an original cause; a motive; these cannot be patented, as no one can claim in either of them an exclusive right."); *Funk Brothers Seed Co. v. Kalo Inoculant Co.*, 333 U.S. 127, 132, 76 USPQ 280, 282 (1948) (combination of six species of bacteria held to be non-statutory subject matter).

[27] The definition of "data structure" is "a physical or logical relationship among data elements, designed to support specific data manipulation functions." The New IEEE Standard Dictionary of Electrical and Electronics Terms 308 (5th ed. 1993).

[28] *Compare In re Lowry*, 32 F.3d 1579, 1583-84, 32 USPQ2d 1031, 1035 (Fed. Cir. 1994) (claim to data structure that increases computer efficiency held statutory) *and Warmerdam*, 33 F.3d at 1360-61, 31 USPQ2d at 1759 (claim to computer having specific memory held statutory product-by-process claim) *with Warmerdam*, 33 F.3d at 1361, 31 USPQ2d at 1760 (claim to a data structure *per se* held non-statutory).

[29] *In re Sarkar*, 588 F.2d 1330, 1333, 200 USPQ 132, 137 (CCPA 1978):

> [E]ach invention must be evaluated as claimed; yet semantogenic considerations preclude a determination based solely on words appearing in the claims. In the final analysis under § 101, the claimed invention, as a whole, must be evaluated for what it is.

Quoted with approval in Abele, 684 F.2d at 907, 214 USPQ at 687. *See also In re Johnson*, 589 F.2d 1070, 1077, 200 USPQ 199, 206 (CCPA 1978) ("form of the claim is often an exercise in drafting").

[30] *See, e.g.*, *Warmerdam*, 33 F.3d at 1361, 31 USPQ2d at 1760 (claim to a data structure *per se* held non-statutory).

[31] Computer Dictionary 210 (Microsoft Press, 2d ed. 1994):

> Data consists of facts, which become information when they are seen in context and convey meaning to people. Computers process data without any understanding of what that data represents.

[32] *See supra* note 29.

[33] *O'Reilly v. Morse*, 56 U.S. (15 How.) at 112-14.

[34] *Id.* at 114-19.

[35] Products may be either machines, manufactures or compositions of matter.

> A machine is:
>
> a concrete thing, consisting of parts or of certain devices and combinations of devices.

Burr v. Duryee, 68 U.S. (1 Wall.) 531, 570 (1863).

> A manufacture is:
>
> the production of articles for use from raw or prepared materials by giving to these materials new forms, qualities, properties or combinations, whether by hand-labor or by machinery.

Diamond v. Chakrabarty, 447 U.S. at 308, 206 USPQ at 196-97 (quoting *American Fruit Growers, Inc. v. Brogdex Co.*, 283 U.S. 1, 11 (1931).

> A composition of matter is:
>
> a composition[] of two or more substances [or] . . . a[] composite article[], whether . . . [it] be the result of chemical union, or of mechanical mixture, whether . . . [it] be [a] gas[], fluid[], powder[], or solid[].

Diamond v. Chakrabarty, 447 U.S. at 308, 206 USPQ at 197 (quoting *Shell Development Co. v. Watson*, 149 F. Supp. 279, 280, 113 USPQ 265, 266 (D.D.C. 1957), *aff'd per curiam*, 252 F.2d 861, 116 USPQ 428 (D.C. Cir. 1958).

[36] *See, e.g.*, *Lowry*, 32 F.3d at 1583, 32 USPQ2d at 1034-35; *Warmerdam*, 33 F.3d at 1361-62, 31 USPQ2d at 1760.

[37] *Cf. In re Iwahashi*, 888 F.2d 1370, 1374-75, 12 USPQ2d 1908, 1911-12 (Fed. Cir. 1989), *cited with approval in Alappat*, 33 F.3d at 1544 n.24, 31 USPQ2d at 1558 n.24.

[38] "Specific software" is defined as a set of instructions implemented in a specific program code segment. *See* Computer Dictionary 78 (Microsoft Press, 2d ed. 1994) for definition of "code segment."

[39] *See Diamond v. Diehr*, 450 U.S. at 183-84, 209 USPQ at 6 (quoting *Cochrane v. Deener*, 94 U.S. 780, 787-88 (1877) ("A [statutory] process is a mode of treatment of certain materials to produce a given result. It is an act, or a series of acts, performed upon the subject-matter to be transformed and reduced to a different state or thing. . . . The process requires that certain things should be done with certain substances, and in a certain

order; but the tools to be used in doing this may be of secondary consequence.").

[40] *See Alappat*, 33 F.3d at 1543, 31 USPQ2d at 1556-57 (quoting *Diamond v. Diehr*, 450 U.S. at 192, 209 USPQ at 10). *See also id.* at 1569, 31 USPQ2d at 1578-79 (Newman, J., concurring) ("unpatentability of the principle does not defeat patentability of its practical applications") (citing *O'Reilly v. Morse*, 56 U.S. (15 How.) at 114-19).

[41] *Diamond v. Diehr*, 450 U.S. at 187, 209 USPQ at 8.

[42] *See In re Gelnovatch*, 595 F.2d 32, 41 n.7, 201 USPQ 136, 145 n.7 (CCPA 1979) (data-gathering step did not measure physical phenomenon).

[43] *Schrader*, 22 F.3d at 294, 30 USPQ2d at 1459 *citing with approval Arrhythmia*, 958 F.2d at 1058-59, 22 USPQ2d at 1037-38; *Abele*, 684 F.2d at 909, 214 USPQ at 688; *In re Taner*, 681 F.2d 787, 790, 214 USPQ 678, 681 (CCPA 1982).

[44] *See supra* note 9.

[45] In *Sarkar*, 588 F.2d at 1335, 200 USPQ at 139, the court explained why this approach must be followed:

> No mathematical equation can be used, as a practical matter, without establishing and substituting values for the variables expressed therein. Substitution of values dictated by the formula has thus been viewed as a form of mathematical step. If the steps of gathering and substituting values were alone sufficient, every mathematical equation, formula, or algorithm having any practical use would be per se subject to patenting as a "process" under § 101. Consideration of whether the

> substitution of specific values is enough to convert the disembodied ideas present in the formula into an embodiment of those ideas, or into an application of the formula, is foreclosed by the current state of the law.

[46] *See supra* note 40.

[47] *See, e.g.*, *In re Bernhart*, 417 F.2d 1395, 1400, 163 USPQ 611, 616 (CCPA 1969).

[48] *Schrader*, 22 F.3d at 293-94, 30 USPQ2d at 1458-59.

[49] *Warmerdam*, 33 F.3d at 1360, 31 USPQ2d at 1759.

[50] *See, e.g.*, *In re Meyer*, 688 F.2d 789, 794-95, 215 USPQ 193, 197 (CCPA 1982) ("Scientific principles, such as the relationship between mass and energy, and laws of nature, such as the acceleration of gravity, namely, a=32 ft./sec.^{2}, can be represented in mathematical format. However, some mathematical algorithms and formulae do not represent scientific principles or laws of nature; they represent ideas or mental processes and are simply logical vehicles for communicating possible solutions to complex problems. The presence of a mathematical algorithm or formula in a claim is merely an indication that a scientific principle, law of nature, idea or mental process may be the subject matter claimed and, thus, justify a rejection of that claim under 35 U.S.C. § 101; but the presence of a mathematical algorithm or formula is only a signpost for further analysis."). *Cf. Alappat*, 33 F.3d at 1543 n.19, 31 USPQ2d at 1556 n.19 in which the Federal Circuit recognized the confusion:

> The Supreme Court has not been clear . . . as to whether such subject matter is excluded from the scope of § 101 because it represents laws of nature, natural phenomena, or abstract ideas. *See Diehr*, 450 U.S. at 186 (viewed

mathematical algorithm as a law of nature); *Benson*, 409 U.S. at 71-72 (treated mathematical algorithm as an "idea"). The Supreme Court also has not been clear as to exactly what kind of mathematical subject matter may not be patented. The Supreme Court has used, among others, the terms "mathematical algorithm," "mathematical formula," and "mathematical equation" to describe types of mathematical subject matter not entitled to patent protection standing alone. The Supreme Court has not set forth, however, any consistent or clear explanation of what it intended by such terms or how these terms are related, if at all.

[51] *Walter*, 618 F.2d at 769, 205 USPQ at 409 (Because none of the claimed steps were explicitly or implicitly limited to their application in seismic prospecting activities, the court held that "[a]lthough the claim preambles relate the claimed invention to the art of seismic prospecting, the claims themselves are not drawn to methods of or apparatus for seismic prospecting; they are drawn to improved mathematical methods for interpreting the results of seismic prospecting."). *Cf. Alappat*, 33 F.3d at 1544, 31 USPQ2d at 1558.

[52] *Walter*, 618 F.2d at 769-70, 205 USPQ at 409.

[53] *See supra* note 45.

[54] *Taner*, 681 F.2d at 788, 214 USPQ at 679.

[55] *Abele*, 684 F.2d at 908, 214 USPQ at 687 ("The specification indicates that such attenuation data is available only when an X-ray beam is produced by a CAT scanner, passed through an object, and detected upon its exit. Only after these steps have been completed is the algorithm performed, and the resultant modified data displayed in the required format.").

[56] *Gelnovatch*, 595 F.2d at 41 n.7, 201 USPQ at 145 n.7 ("Appellants' claimed step of perturbing the values of a set of process inputs (step 3), in addition to being a mathematical operation, appears to be a data-gathering step of the type we have held insufficient to change a nonstatutory method of calculation into a statutory process. . . . In this instance, the perturbed process inputs are not even measured values of physical phenomena, but are instead derived by numerically changing the values in the previous set of process inputs.").

[57] *Sarkar*, 588 F.2d at 1331, 200 USPQ at 135.

[58] *See Sarkar*, 588 F.2d at 1332 n.6, 200 USPQ at 136 n.6 ("post-solution" construction that was being modeled by the mathematical process not considered in deciding § 101 question because applicant indicated that such construction was not a material element of the invention).

[59] *Parker v. Flook*, 437 U.S. 584, 585, 198 USPQ 193, 195 (1978).

[60] *Walter*, 618 F.2d at 770, 205 USPQ at 409 ("If § 101 could be satisfied by the mere recordation of the results of a nonstatutory process on some record medium, even the most unskilled patent draftsman could provide for such a step.").

[61] *Gelnovatch*, 595 F.2d at 41 n.7, 201 USPQ at 145 n.7.

[62] *Abele*, 684 F.2d at 909, 214 USPQ at 688 ("This claim presents no more than the calculation of a number and display of the result, albeit in a particular format. The specification provides no greater meaning to 'data in a field' than a matrix of numbers regardless of by what method generated. Thus, the algorithm is neither explicitly nor implicitly applied to any certain process. Moreover, that the result is displayed as a shade of gray rather

than as simply a number provides no greater or better information, considering the broad range of applications encompassed by the claim.").

[63] *In re De Castelet*, 562 F.2d 1236, 1244, 195 USPQ 439, 446 (CCPA 1977) ("That the computer is instructed to transmit electrical signals, representing the results of its calculations, does not constitute the type of 'post solution activity' found in *Flook*, [437 U.S. 584, 198 USPQ 193 (1978)], and does not transform the claim into one for a process merely *using* an algorithm. The final transmitting step constitutes nothing more than reading out the result of the calculations.").

[64] *E.g.*, *Warmerdam*, 33 F.3d at 1360, 31 USPQ2d at 1759. *See also Schrader*, 22 F.3d at 295, 30 USPQ2d at 1459.

[65] *See supra* note 18 and accompanying text.

[66] Computer Dictionary 353 (Microsoft Press, 2d ed. 1994) (definition of "self-documenting code").

[67] *See In re Barker*, 559 F.2d 588, 591, 194 USPQ 470, 472 (CCPA 1977), *cert. denied*, *Barker v. Parker*, 434 U.S. 1064 (1978) (a specification may be sufficient to enable one skilled in the art to make and use the invention, but still fail to comply with the written description requirement). *See also In re DiLeone*, 436 F.2d 1404, 1405, 168 USPQ 592, 593 (CCPA 1971).

[68] *See, e.g.*, *Northern Telecom v. Datapoint Corp.*, 908 F.2d 931, 941-43, 15 USPQ2d 1321, 1328-30 (Fed. Cir.), *cert. denied*, *Datapoint Corp. v. Northern Telecom*, 498 U.S. 920 (1990) (judgment of invalidity reversed for clear error where expert testimony on both sides showed that a programmer of reasonable skill could write a satisfactory program with ordinary effort based on the disclosure); *DeGeorge v. Bernier*, 768 F.2d 1318, 1324,

226 USPQ 758, 762-63 (Fed. Cir. 1985) (superseded by statute with respect to issues not relevant here) (invention was adequately disclosed for purposes of enablement even though all of the circuitry of a word processor was not disclosed, since the undisclosed circuitry was deemed inconsequential because it did not pertain to the claimed circuit); *In re Phillips*, 608 F.2d 879, 882-83, 203 USPQ 971, 975 (CCPA 1979) (computerized method of generating printed architectural specifications dependent on use of glossary of predefined standard phrases and error-checking feature enabled by overall disclosure generally defining errors); *In re Donohue*, 550 F.2d 1269, 1271, 193 USPQ 136, 137 (CCPA 1977) ("Employment of block diagrams and descriptions of their functions is not fatal under 35 U.S.C. § 112, first paragraph, providing the represented structure is conventional and can be determined without undue experimentation."); *In re Knowlton*, 481 F.2d 1357, 1366-68, 178 USPQ 486, 493-94 (CCPA 1973) (examiner's contention that a software invention needed a detailed description of all the circuitry in the complete hardware system reversed).

[69] *See In re Naquin*, 398 F.2d 863, 866, 158 USPQ 317, 319 (CCPA 1968) ("When an invention, in its different aspects, involves distinct arts, that specification is adequate which enables the adepts of each art, those who have the best chance of being enabled, to carry out the aspect proper to their specialty."); *Ex parte Zechnall*, 194 USPQ 461, 461 (Bd. App. 1973) ("appellants' disclosure must be held sufficient if it would enable a person skilled in the electronic computer art, in cooperation with a person skilled in the fuel injection art, to make and use appellants' invention").

[70] *See In re Scarbrough*, 500 F.2d 560, 565, 182 USPQ 298, 301-02 (CCPA 1974) ("It is not enough that a person skilled in the art, by carrying on investigations along the line indicated in the instant application, and by a great amount of work eventually

might find out how to make and use the instant invention. The statute requires the application itself to inform, not to direct others to find out for themselves (citation omitted)."); *Knowlton*, 481 F.2d at 1367, 178 USPQ at 493 (disclosure must constitute more than a "sketchy explanation of flow diagrams or a bare group of program listings together with a reference to a proprietary computer on which they might be run"). *See also In re Gunn*, 537 F.2d 1123, 1127-28, 190 USPQ 402, 405 (CCPA 1976); *In re Brandstadter*, 484 F.2d 1395, 1406-07, 17 USPQ 286, 294 (CCPA 1973); and *In re Ghiron*, 442 F.2d 985, 991, 169 USPQ 723, 727-28 (CCPA 1971).

[71] *Cf. In re Gulack*, 703 F.2d 1381, 1385, 217 USPQ 401, 404 (Fed. Cir. 1983) (when descriptive material is not functionally related to the substrate, the descriptive material will not distinguish the invention from the prior art in terms of patentability).

Annex for Appendix 3

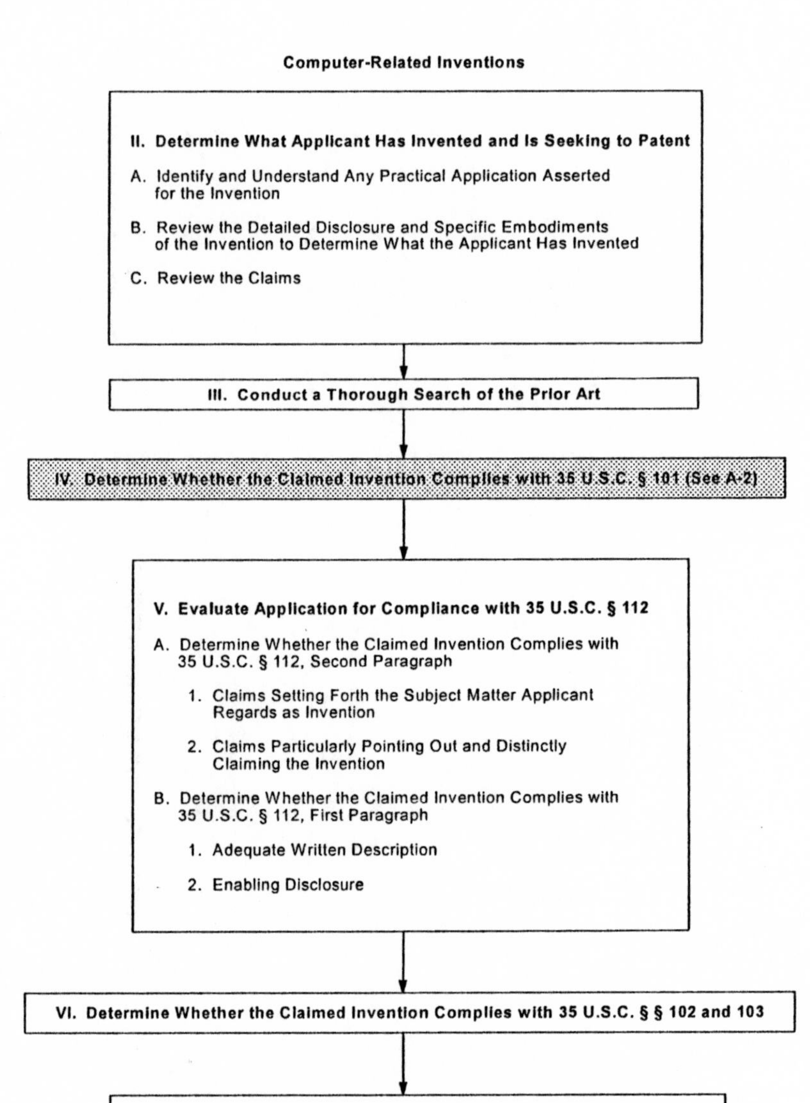

Computer-Related Inventions
II. Determine What Applicant Has Invented and Is Seeking to Patent
A. Identify and Understand Any Practical Application Asserted for the Invention
B. Review the Detailed Disclosure and Specific Embodiments of the Invention to Determine What the Applicant Has Invented
C. Review the Claims
III. Conduct a Thorough Search of the Prior Art
IV. Determine Whether the Claimed Invention Complies with 35 U.S.C. § 101 (See A-2)
V. Evaluate Application for Compliance with 35 U.S.C. § 112
A. Determine Whether the Claimed Invention Complies with 35 U.S.C. § 112, Second Paragraph
1. Claims Setting Forth the Subject Matter Applicant Regards as Invention
2. Claims Particularly Pointing Out and Distinctly Claiming the Invention
B. Determine Whether the Claimed Invention Complies with 35 U.S.C. § 112, First Paragraph
1. Adequate Written Description
2. Enabling Disclosure
VI. Determine Whether the Claimed Invention Complies with 35 U.S.C. § § 102 and 103
VII. Clearly Communicate Findings, Conclusions and Their Bases

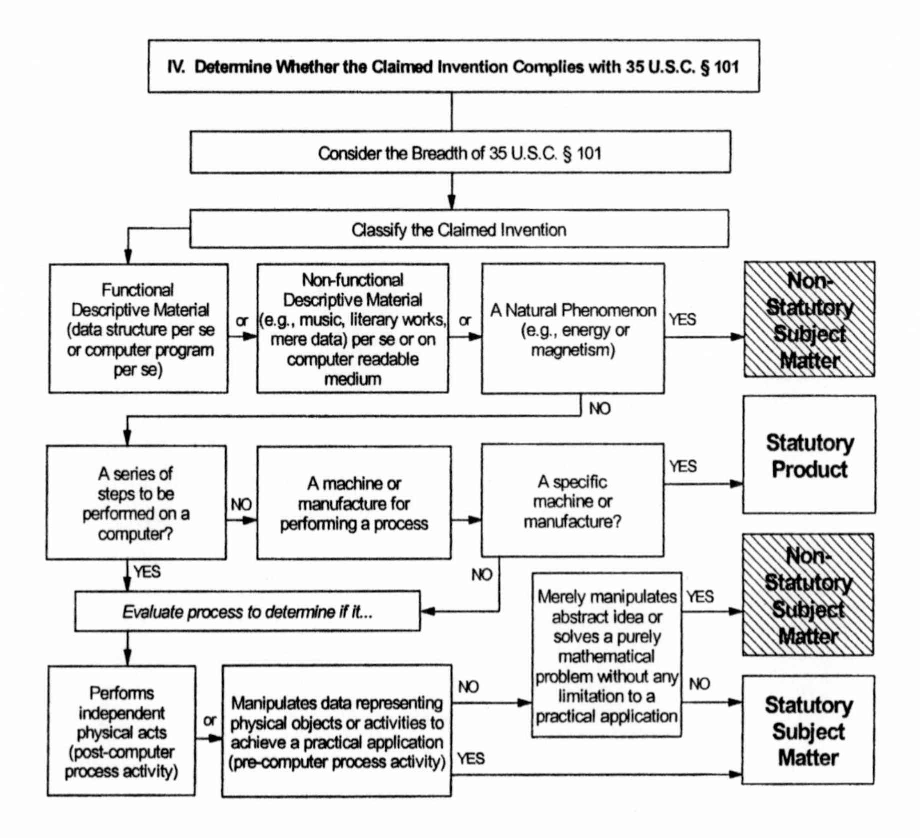
IV. Determine Whether the Claimed Invention Complies with 35 U.S.C. § 101
Consider the Breadth of 35 U.S.C. § 101
Classify the Claimed Invention
Functional Descriptive Material (data structure per se or computer program per se)
or
Non-functional Descriptive Material (e.g., music, literary works, mere data) per se or on computer readable medium
or
A Natural Phenomenon (e.g., energy or magnetism)
YES
Non-Statutory Subject Matter
NO
A series of steps to be performed on a computer?
NO
A machine or manufacture for performing a process
A specific machine or manufacture?
YES
Statutory Product
YES
NO
Evaluate process to determine if it...
Merely manipulates abstract idea or solves a purely mathematical problem without any limitation to a practical application
YES
Non-Statutory Subject Matter
Performs independent physical acts (post-computer process activity)
or
Manipulates data representing physical objects or activities to achieve a practical application (pre-computer process activity)
NO
NO
YES
Statutory Subject Matter

Table of Authorities

1. CASES

2. *U.S. CONSTITUTION AND U. S. CODE*

3. CODE OF FEDERAL REGULATIONS, FEDERAL REGISTER, AND PATENT OFFICE GAZETTE

4. LEGISLATION

5. *UNIFORM COMMERCIAL CODE*

6. *U.S. PATENTS*

7. MISCELLANEOUS

Index

Printed in the United States
18455LVS00002BA/1-24